Educational Change towards Problem Based Learning: An Organizational Perspective

Innovation and Change in Education — A Cross-cultural Perspective

Volume 5

Series Editor

Prof. Dr. Xiangyun Du
Aalborg University, Denmark

Nowadays, educational institutions are being challenged when professional competences and expertise become progressively more complex. This is mainly because problems are more technology-bounded, unstable and ill-defined with the involvement of various integrated issues. To solve these problems, it requires interdisciplinary knowledge, collaboration skills, innovative thinking among other competences. In order to facilitate students with the competences expected in professions, educational institutions worldwide are implementing innovations and changes in many aspects.

This book series includes a list of research projects that document innovation and change in education. The topics range from organizational change, curriculum design and innovation, pedagogy development, to the role of teaching staff in the educational change process, and quality issues, among others. A cross-cultural perspective is studied in this book series that includes two layers. First, research contexts in these books include different countries with various educational traditions, systems and societal backgrounds. Second, the impact of professional and institutional cultures such as engineering, medicine and health, and teachers' education are also taken into consideration in these research projects.

For a list of other books in this series, visit www.riverpublishers.com

Educational Change towards Problem Based Learning: An Organizational Perspective

Huichun Li

River Publishers
Aalborg

Published, sold and distributed by:
River Publishers
Niels Jernes Vej 10
9220 Aalborg Ø
Denmark
www.riverpublishers.com

ISBN: 978-87-92982-70-4

Contents

Preface

I have been educated in a very traditional lecture-based (sometimes also refers to subject-based) education atmosphere in China. Along my primary school, secondary school, and even university, I was exposed to various subjects that I was instructed to memorize the knowledge content that the teachers delivered to me. I was a good learner during that time and I could obtain very good remarks during the examination. Everything seemed fine until one day I became frustrated when I realized that I could barely recall many knowledge materials that I had once memorized only several weeks after my examination.

At that moment I had already become an education researcher at university. Through intensive studies on education, I came to realize that traditional lecture-based educational approach is not perfect but with many drawbacks in terms of suppressing student learning motivation and subjectivity, overemphasizing knowledge retention, ignoring skill development, etc. Fortunately, our life is always progressing: we encounter many problems bothering us, but we also develop new approaches to solving the problems. Over the past several decades, many innovations have been emerged within educational domain with the attempt of remedying the problem of the traditional lecture-based educational approach, promoting student learning motivation, and thus improving the overall educational quality. Problem-based learning (PBL) approach is one of the typical examples of such educational innovations. The intention of PBL is to formulate a new way of educating students in which both learning and curriculum design starts from a problem in real life, students work in the form of groups on the problem, students have large autonomy to direct their learning process, teachers serve as facilitators rather than instructors. In practice, PBL has brought many benefits for students, teachers, and education institutions (Strobel & van Barneveld, 2009), and therefore it has been adopted by an increasing number of education institutions worldwide.

However, no educational change can be completed overnight, and so is PBL. Through literature review, I noticed that the change to PBL is not always easy for education institutions, especially for those with long history of traditional lecture-based learning tradition. Many education institutions in the change to PBL are challenged by various issues in terms of resource, curriculum design, management, staff motivation, and so on. So, where are all these challenges coming from when we making the change to PBL and how should we deal with them? It is this question that leads to the production of this book.

Many studies have been made on the challenges for the change to PBL, and educational theorists and practitioners have already put forward suggestions from various perspectives on how to facilitate such an educational change (Savin-Baden, 2000, 2003; Moesby, 2004; Kolmos & De Graaff, 2007). These suggestions are valuable; however, methodologically, offering prescriptions does not necessarily secure the improvement of the practice since there might be a lack in genuinely understanding reality. The ultimate goal of educational research is to improve practice. In order to accomplish this goal, it is necessary to start from touching and understanding reality rather than making prescriptions. The reality is not purely objective; rather, it is constructed by people who engage in everyday practice (Henriksen et al., 2004). A genuine understanding of reality can be better reached when we document and analyze how people interpret the world. Therefore, this book is primarily concerned with how people (managers, teaching and research staff, students, etc.) understand and interpret PBL and the relevant issues when education institutions make the change to PBL.

This book is particularly concerned with the change to PBL in the domain of higher education. Two universities from different national contexts are included in this study. Both of them are in the process of introducing PBL elements and implementing PBL curriculum throughout the entire university during the last few years. Their change to PBL is not smooth but replete with many challenges in terms of staff's diverse interpretation of PBL, unsupportive university policies, internal tension, etc. A genuine documentation of the change process has been made as thorough as possible so as to demonstrate what really happens in reality as much as possible. By an intensive documentation and analysis of how people interpret PBL, this study shows that the conceptions and perceptions of the educational practitioners regarding PBL implementation at the selected universities could be responsible for the challenges for the change to PBL to a large extent.

A further concern when dealing with these challenges for PBL implementation is the change approach, such as how to manage the change process. So far, a widely accepted assumption among many researchers is that there might be a best way to manage the change process for a particular context. Even the contingent theorists may also support this assumption, although they admit that the best change approach is highly dependent on the context. However, this book tends to refute this popular assumption by arguing that all change approaches have their own benefits and risks. As later shown in this book, both normative approaches (e.g. managers enforce directions) and non-normative approaches (e.g. managers allow staff's individual exploration) have their own strengths and weakness. Therefore, what matters is not to identify a best way to implement PBL but to make choices among different change approaches with different benefits and risks.

Finally, I hope this book could be useful for educational researchers, managers, teachers and anyone who is interested in education research. This book might not be directly giving suggestions or guidelines to direct their practice, but it can provide a new perspective to help them better understand reality and make further reflections on their own thoughts and practices.

Acknowledgements

Many thanks are due to Lars Bo Henriksen and Xiangyun Du from Aalborg University, who, over the last few years, have profoundly influenced me by challenging my thinking, offering me constructive comments, as well as encouraging me to write. I would like to give my sincere gratitude to Alex Stojcevski, Cagil Ozansoy, Baozhi Sun and Qiang Liu who assist me in gaining access to the universities involved in this study. I am profoundly in debt to all participants who have been involved in my interviews and observations. Although their names cannot be listed here, but definitely this book is impossible to be completed without their support. I am also grateful to all my colleagues at Aalborg University for their valuable comments and suggestions along the last few years. Further, I am sincerely grateful to China Scholarship Council and Aalborg University, who made it financially possible for me to finish this manuscript. Finally, special thanks are given to Sarah Freire-Gibb for her careful proofreading on this book.

List of Abbreviations

PBL: Problem Based Learning
LBL: Lecture Based Learning
XU: X University in Australia
EE: School of Electric and Electronic Engineering at X University
ACME: School of Architecture, Civil and Mechanic Engineering at X University
ZU: Z University in China
CBMS: College of Basic Medical Science at Z University
AAU: Aalborg University in Denmark

List of Figures

1

Introduction

1.1 Critiques of Traditional Educational Approach

Our discussions on education start from a quotation from a popular Chinese blog:

> "Three years ago, I could do trigonometric functions and simultaneous equations. I could recite periodic tables, I knew oxidation-reduction reactions and Newton's three laws, I could draw atmospheric circulation map, and I knew that plant cells have walls but animal cells do not, I could recite several ancient classical poems and essays...... Now I have forgotten all of them. I am completely illiterate!"
>
> (Thinktask, 2011)

The monolog above depicts a young man's frustration with his educational experience after his graduation from high school. It can be inferred that the educational system taught him a wealth of information from a wide range of disciplines. While he had memorized all the required content in the short term, he could barely recall any of it several years after graduation, meaning long-term retention was lacking. Therefore, in the long run, the educational approach described by the writer fails to accomplish its goal of knowledge retention.

If we quote Dewey's (1938) term of continuity, which he sets as a crucial dimension of the educational experience, it can be noted that the educational experience described above did not contribute to the learner's current knowledge level. Furthermore, the writer's current life does not need the knowledge from the past, explaining why the student has forgotten a large amount of the information he had acquired. It can be inferred that knowledge retention is widely regarded as the main objective of the educational system since the monolog tends to equate the lack of knowledge retention to being "completely illiterate." The writer is frustrated by this fact and attaches crucial meaning to knowledge retention in education. The value of knowledge retention in education is taken for granted in a system using a traditional approach to education. Thus, education is regarded as losing its value if it fails to help students retain knowledge in the long run. However, is the essence of education merely knowledge acquisition and retention? How should we understand the value of education?

This brings us to consider and examine the features of the traditional educational approach, or what is widely labeled as LBL (Lecture-Based Learning). Within the traditional educational context, the focal point of education is on the acquisition of knowledge: learning objectives, in most cases, highlight the importance of knowledge retention, and the assessment becomes a process of examining the memorization ability of the individual student. Students are expected to successfully repeat, reproduce and recall the content that they have acquired from lectures. Along with an emphasis on knowledge retention, a traditional educational approach is designed to facilitate knowledge transmission. Teachers are authorities and representatives of knowledge and traditional wisdom, and they transfer the intended body of information and tradition into the minds of students through the use of lecturing, a one-way delivery method. In this learning environment, teachers dominate the teaching and learning process in the classroom where the students are simply required to behave as passive learners whose main task is to memorize content.

This LBL approach is methodologically problematic in attaining its traditional goal of knowledge retention, referred to the monolog at the beginning of this section. Therefore, a paradox emerges since LBL ironically does not function well in realizing its intended objectives, even if it may consolidate the memorization of knowledge.

More problematic still is that the traditional educational approach tends to reproduce tradition. A traditional educational approach (LBL) is described as the banking model (Freire, 1970). Students are treated as empty accounts where teachers deposit information through a process of lecturing and then retrieve the information when assessing students' learning outcomes. In this sense, the mind of the student is simply treated as a medium in which to reproduce the information deposited by the teacher. The problem with a traditional educational approach, among many other things, is that it directs the student towards the past rather than the present or the future.

John Dewey (1916) identified two contradictory educational thoughts: retrospective education and prospective education. Retrospective education tends to see education as constantly reproducing tradition and thus accommodates the future through the past. A retrospective view of education relies heavily upon past experience and tradition, encourages the reproduction of tradition and sees the future as a repetition of the past. This viewpoint may have made sense in a pre-industrial society where societal change was quite limited and where several generations lived in the same manner and shared common problems. Therefore, by learning retrospective knowledge, students were able to adapt to society and use the knowledge they had gained to solve present problems. As long as society remained stable, knowledge from the past could still be an effective tool to solve present problems. Nevertheless, in a modern society, traditional wisdom is, to some extent, no longer useful when dealing with the problems of today since society has changed. Thus, we need a new way of conceiving of education as facilitating human beings' survival and adaptation in the world.

Prospective education, on the other hand, directs students to current and future situations which should be treated as an opportunity for constant renewal and a process of continuous self-growth. Reproducing the past is not stressed in this view. Furthermore, it refutes the traditional educational approach which consists of externally imposed knowledge and lacks connection to the learners' personal experience, and proposes that the educational design should take into consideration both students' personal experiences and their external environment.

Dewey(1916) noted the shortcomings of the traditional educational approach in preparing students for the present and the future. However, what constituted the present and the future in the contemporary world remained unaddressed at Dewey's time. To elucidate on the traits of a contemporary society we need contemporary wisdom. Bauman described the present social patterns as a "liquid society" (Bauman, 2000) where the traditionally solid, fixed social institutions are gradually disintegrating and being replaced by "fluidity" in terms of interpersonal relationships, working patterns, communities, etc. The human beings living in our current society have become "consumers", "tourists" or "hunters" who have no ultimate goal to pursue but who are always "on the move and bound to remain so"(Bauman, 1998, p.78). Life is "continuing and continuous hunting" (Bauman, 2005, p. 311), an endless process of disposing of the old and hunting for the new. Within such a societal setting, retrospective education, which aims to reproduce a stable tradition, cannot function well. If traditional education aims to "building upon a firm foundation, adding new knowledge to the already acquired volume and thereby enabling the pursuit of the selected trajectory" (Bauman, 2005, p.312), then the education in a liquid society is "an unending succession of new beginnings, moved more by a swift forgetting of the previously acquired knowledge than by an acquisition of new knowledge; it militated against retention and memorizing" (Bauman, 2005, p.313).

One implication of Bauman's argument[1] is that knowledge in contemporary society should somehow be treated as having only a short life span due to the fast emergence of new knowledge and the even faster abandonment and disposal of old knowledge. Since the nature of knowledge has changed from long-term retention to information needed only in the short term, it is no longer necessary to retain the traditional education aim of pursuing long-term knowledge retention. The traditional education approach for knowledge retention and reproduction is neither necessary nor possible in a modern world since knowledge is rapidly discarded, not just because of the utility of the knowledge, but also due to the volatility and the fluidity of modern individual lives. As a result, education should cultivate students' ability of "hunting" for the new and the "hunting" skills in general.

1.2 PBL as a Solution

Various educational innovations have emerged to facilitate students' self-renewal and foster their renewal of skills. Among the new approaches to education is PBL (Problem-Based Learning) with a history of over four decades in higher education. PBL originated in the medical field at McMaster University Medical School in Canada in the late1960s as an alternative to the traditional education approach (Savin-Baden & Major, 2004). To replace the traditional lectures, McMaster University designed a series of simulated patient problems that students solved by working in small teams. As a result, the teacher served as a facilitator, rather than the knowledge authority who directed the students' learning process. It was documented that the students of problem-based learning tended to have higher motivation, better problem-solving skills, improved group-work skills and increased self-study skills (Dolmans and Schmidt, 1996; Dochy, Segers, Bossche, & Gijbels, 2000; Bowe, Brian & Cowan, 2004; Strobel & van Barneveld, 2009), when comparing with their non-PBL counter parts. Apart from McMaster University, PBL was also adopted in other fields as a replacement for LBL. In the early 1970s, two newly founded universities, Aalborg University and Roskilde University

[1]Bauman's view of knowledge seems a bit extreme. Here we should consider Aristotle's differentiation of knowledge: episteme, techne and phronesis. We should be aware that not all knowledge types could be treated in the same manner as Bauman's.

in Denmark, which featured engineering sciences, and humanities and social sciences, respectively, decided to employ PBL as the fundamental educational approach throughout the entire university (Kolmos, Fink & Krog, 2004). PBL proved to be quite promising. After its success in the early phases, PBL has gradually spread to an increasing number of educational fields worldwide.

Though PBL can find its theoretical roots in a variety of domains (Savin-Baden & Major, 2004), the most influential theory for PBL is constructivism (e.g. Piaget, 1953; Vygotsky, 1962) and the educational thoughts developed by Dewey (1916, 1938). Knowledge is not an absolute objective matter transmitted to the students' minds and disconnected from the students' experiences; rather, it is constructed and formulated through learners' engagement in learning activities, interaction with previous experiences and their surrounding environment.

Although there are various definitions of PBL (Barrows & Tamblyn, 1980; Boud, 1985; Graaff & Kolmos, 2003), theoretically, three major principles of PBL can be identified: problem centeredness, interdisciplinary learning and social learning (Kolmos, 2008). Problem centeredness refers to the priority of the problem in the whole learning process. Learning starts with a real problem that later dominates the learning process. A problem can take various forms, a social discrepancy, a theoretical dilemma, an engineering project or a medical case (Kolmos, Fink & Krogh, 2004). Interdisciplinary learning implies an approach that coordinates and integrates different disciplinary fields in contrast to LBL, which is mainly based upon one single subject. PBL students are encouraged to develop the learning process and solve their problem through the use of theories and methods from different disciplinary fields since problems in real life are often complex matters and cannot be readily solved by a single disciplinary method. Finally, students are required to work in groups to increase interaction between students. By these principles, it can be noted that PBL links students to real-life situations by working on real problems and to social contexts through the use of teamwork. An implicit part of PBL is student centeredness, sometimes termed self-directed or participant-directed learning (De Graaff & Kolmos, 2003), which implies that students dominate the learning process. This gives students rights to decide their learning objectives, content and activities, to a large extent. Although PBL does not completely expel lecturing and tutorials from the learning process, on the whole, the teacher, in most cases, appears as a facilitator or a consultant instead of a knowledge authority.

By specifying PBL principles, it can be noted that PBL differs in many ways from LBL in terms of how to organize educational activities. LBL relies upon subject-based knowledge, single discipline-based learning, individual learning, and the teachers' dominance in the learning process, which transform education into a one-way imparting of knowledge and a reproduction of tradition. In contrast, PBL stimulates student engagement in the present, in terms of working on real-life problems in the form of teamwork, and therefore, propels student self-growth and constant knowledge renewal.

1.3 The Value of PBL

The merits of PBL has been widely documented in terms of promoting students' learning outcomes, motivating students to learn, and fostering students' learning habits, etc. (e.g. Dolmans & Schmidt, 1996; Dochy, Segers, Bossche & Gijbels, 2000; Bowe, Brian & Cowan, 2004; Strobel & van Barneveld, 2009). Specifically, a meta synthesis (Bair, 1999) concerning the learning outcome comparison between the traditional approach and the PBL approach revealed that the students from PBL contexts significantly outperform traditionally taught students in terms of long term knowledge retention, skill-oriented performance, and mixed development of both knowledge and skills (Strobel & van Barneveld, 2009). Short-term knowledge acquisition and retention seems to favor the traditional approach; however, the measurement of short-term gains for PBL is methodologically problematic since the essence of PBL is preparing for long-term information transfer rather than short-term wins (Barrow, 2002). Moreover, Savin-Baden argues that the merits of PBL to students, staff members and employers, have been underestimated to a large extent (Savin-Baden, 2000:26). In addition to learning outcomes, PBL is also advantageous in dealing with negative student attitudes towards learning, students' motivation in daily performance and attrition issue. For instance, there is research documenting that students from PBL contexts have a more positive attitude towards their learning environment than their counterparts in traditional programs (De Vries, Schmidt & de Graaff, 1989). Moreover, PBL students more frequently use textbooks and engage in informal discussions with peers than non-PBL students who prefer to rely on lecture notes (Blumberg & Michael, 1992).

Furthermore, PBL carries far more merits than these instrumental values. The significance of PBL is closely linked to our perception of knowledge. If we quote Aristotle's knowledge taxonomy as episteme, techne and phronesis (Flyvbjerg, 2001; Henrisken et al., 2004), it can be concluded that the traditional educational paradigm focuses on the acquisition of epistemic knowledge, which is classified as being context independent and is thus weak in promoting students' acquisition of both techne and phronesis knowledge. PBL involves the presence of intense activities which encourage students to apply knowledge to specific situations and thus facilitate the acquisition of techne knowledge. Most importantly, the application of knowledge along with the intense interaction with other learners encourages students to consider the moral respect of the knowledge; that is, how to use the knowledge. Here, the application of PBL should not be understood in a technical sense, but also in an ethical manner.

Through its application, PBL is able to make education more productive. Being productive here does not refer to the economic sense of promoting effectiveness, but to the sense of producing something new or creating new possibilities, which in turn transforms the learner as well as the tradition (Gallagher, 1992, p. 189). From the perspective of hermeneutics, education is always productive in terms of producing something new and transforming tradition. As for PBL, learning processes become more productive because the approach contains elements of application, which are not limited to a specific utility meaning, but refer to the intention of making knowledge relevant to oneself. This process of "making relevant" creates something new, which may eventually lead to a transformation of tradition.

1.4 Changing to PBL

Given the merits of PBL, a number of higher educational institutes have adopted PBL partly or wholly in their educational programs among various disciplines (see Table 1.1). In these cases, the educational practitioners introduced PBL into the existing higher education systems using a diverse set of strategies and approaches. Some universities adopted PBL only as an instructional strategy, which gave PBL a role in a limited number of subjects. Other universities introduced PBL at a systematic level, meaning the whole curriculum system was redesigned and restructured. This introduction at the foundational level of the university can be indicative of a possible cultural change in its education philosophy.

The process of transitioning to PBL is not an easy task for higher educational institutions, especially for those who bear a long history of LBL and implement PBL as a whole curriculum approach, rather than an instructional strategy. A variety of challenges and obstacles are documented when higher

Table 1.1 Universities that have implemented PBL

Continent	Country/Region	Name of Educational Institutions	Author(s)
Africa	Uganda	Makerere University	Kiguli-Malwadde, E. et al. (2006)
	South Africa	University of Transkei	Meel, B. (2003)
Asia	China(mainland)	Wuhan University	Wang, G. et al. (2008)
	China(Hong Kong)	University of Hong Kong	MacKinnon, M. (1999)
	China(Taiwan)	Kaohsiung Medical University	Lin,Y. et al. (2009)
	Pakistan	Ziauddin Medical University	Nooruddin, Z. & Jaffery, M. (2004)
Australia	Australia	University of Melbourne	Grkovic, I. (2005) Keppell, M. et al. (1998)
		Victoria University	Stojcevski, A., & Ozansoy, C. (2009)
		Newcastle Faculty of Architecture	De Graaff, E. & Cowdroy, R. (1997)
Europe	Switzerland	University of Geneva	Vu, N.et al. (1997)
	Finland	University of Helsinki	Lonka, K. (2000) Lonka, K. (2001)
	Belgium	University of Louvain	Raucent, B. (2001)
	Portugal	Aveiro University	Alarcao, I. (2007) De Oliveira, J. (2007)
	Ireland	Dublin Institute of Technology	Bowe, B. (2007)
	Denmark	Case A	Kolmos, A. (2008)
		Case B (together with case A)	Kolmos, A. (2008)
		Aalborg University, Esbjerg	Kolmos, A. (2002) Kolmos, A. & Algreen-Ussing, H. (2001)
	Netherland	Delft Faculty of Architecture	De Graaff, E. & Cowdroy, R. (1997)
North America	Canada	University of Sherbrooke	Marchais, J. (1991) Maison, P. and Marchais, J.(1991) Marchais, J. (1993)
South America	Argentina	Rosario University of Medicine	Carrera, L. et al. (2003)

educational institutions implementing PBL. At Aveiro University (Alarcao, 2007; De Oliveira, 2007), staff members are quite accustomed to the traditional teaching method and are reluctant to use the new PBL approach in the instructional process. Further, there was a sense of lack in high quality "problems" from industry when the staff members designed the curriculum. At the Dublin Institute of Technology in Ireland (Bowe, 2007), the dominant pedagogical concern still focuses on knowledge coverage, rather than how to use the knowledge. This way of thinking is a significant detriment to PBL implementation. Another practical problem that PBL implementation faces is that higher education is becoming increasingly research oriented, which leads to growing staff interest in research rather than teaching. Staff members are quite reluctant to get involved in teaching activities if their attention is preoccupied by research. Therefore, the implementation of PBL requires the adjustment of teacher evaluation and reward systems at the university. At the Delft Faculty of Architecture in the Netherlands, the debate over the significance of PBL is notably endless. It can be observed that there are a large group of staff members who are opposed to the idea of PBL implementation in their department and at the university as a whole. Some teachers find a conflict of roles in PBL teaching since it requires staff members to transform their role from dominating the teaching process to facilitating the learning process. PBL also has some opponents in the student body who are used to a traditional learning approach and find themselves uncomfortable in a PBL environment. One possible reason for the resistance to PBL at Delft is that the change preparation phase is very short (De Graaff & Cowdroy, 1997). Without substantial preparation, changing to PBL becomes quite chaotic. In the Newcastle Faculty of Architecture in Australia, both staff and students suffer considerably from the workload created by PBL, which leads the staff to choose to "quit" PBL (De Graaff & Cowdroy, 1997). This example from Newcastle should serve as a reminder to PBL practitioners of the necessity of carefully examining the workload PBL will create and refining it to a tolerable scale. At Rosario University of Medicine, the limited human and financial resources greatly restrict the implementation of PBL since PBL calls for substantial amounts of resources in terms of money, manpower and time (Carrera, 2003). Wuhan University in China has a historical background in a traditional educational system which leads staff members to consider PBL to be time consuming and difficult to implement (Wang et al., 2008). In Kaohsiung Medical University in Taiwan, the university experiences a lack of tutors in tutorial when introducing PBL. Also, the training for staff members is far from sufficient (Lin et al., 2009). I will not make a thorough exploration of all the challenges and obstacles PBL implementation faces, but will specify some of them:

- Pedagogical debate on PBL: knowledge coverage and retention or knowledge application
- Staff and student attitudes towards PBL: supportive or opposed
- Resource limitations: human, facility, time, finance
- PBL design: type, size of problem to solve, assessment method
- PBL's incompatibility with a traditional education system: student learning style, teachers' teaching tradition, student assessment methods, teacher evaluation systems
- Problems produced by PBL: workload
- Strategy of change: preparation phase, top down strategy.

This list details some of the difficulties PBL implementation experiences at higher education institutions. However, it is not sufficient to specify the challenges in implementing PBL; rather, it is necessary to examine the methodological flaws of the traditional theories dealing with PBL implementation.

1.5 Revisiting Current PBL Implementation Research

Through an investigation of PBL effectiveness and implementation, Strobel and Barneveld (2009) argue that the effectiveness of PBL has been well documented in a number of studies. However, the research focus should be shifted toward the drivers and challenges of PBL implementation as data is lacking in that area. Savin-Baden (2000) expressed her frustration with current PBL research when she wrote that researchers were attempting to "offer the world of higher education guidance in, and examples of, the implementation of problem based learning but little, in real terms, that deals with the difficulties and complexities of the approach" (p.5). The implication is that the challenges of realizing PBL have not been fully addressed. Following this, this book will focus on the challenges and obstacles universities face when they are in the process of transforming their traditional educational approaches, such as LBL, to PBL. Moreover, this book concerns the following methodological issues when gathering data relevant to PBL implementation.

One flaw in the current PBL studies is that a significant number of studies equate PBL to an instructional approach and they tend to reduce PBL implementation to just a pedagogical matter. Margetson

(1991) emphasized that PBL may be seen as "a conception of knowledge, understanding, and education profoundly different from the more usual conception underlying subject-based learning" (p.43). Biggers and Tang (2007) built on this thought when they proposed that educational changes should take place in a systematic manner entailing three elements: the learning objective, the teaching and learning method and the assessment. Changes to one of the three elements require adjustments to the other two. Furthermore, Kolmos and De Graaff (2007) link PBL implementation to changes in not only curriculum, but also in larger organizational elements such as physical resources, infrastructure and the culture of the organization. Consequently, universities wanting to introduce PBL should expect changes to the organization as a whole and to adjust curriculum during a successful transition (Kolmos & Algreen-Ussing, 2001). Organizational change is key. A complete shift toward PBL requires an entire institutional change from assessments to decision-making structures and organizational culture (Moesby, 2004). The resulting cultural changes are especially important for successful PBL implementation (Kolmos, 2002). Adding complexity to PBL implementation is that educational changes at a university are largely conditioned by local, national and international contexts. Therefore, the adoption of a systematic perspective and the exploration of the organizational, societal and cultural dimensions of the university need to be considered when dealing with implementing PBL.

A further weakness regarding current research studies of PBL is the focus on design and implementation instead of the organizational members' understandings and interpretations of PBL at higher education institutions. Further, the voices of the students and the staff are largely missing from the studies on PBL (Savin-Baden, 2000). We should not neglect the human factor of the organization since it has a significant impact on the success of change on all levels of the organization (Henriksen et al., 2004). There are many studies that document the relationship between the teachers' conceptions of teaching in relation to their instructional approaches (Trigwell & Prosser, 1996; Trigwell, Prosser & Waterhouse, 1999), and the relationship between the students' conceptions of learning and their learning approaches (Ramsden, 1992; Turner, 1995; Savin-Baden, 2000). Therefore, in order to attain a genuine understanding of the organizational changes when PBL is implemented, it is necessary to look at the organizational members' understanding of the process.

1.6 Book Outline

Given that the challenges and obstacles for PBL implementation have not been fully addressed yet, the research questions are: What are the challenges and obstacles to an organization when it shifts from its traditional educational paradigm to PBL? Moreover, how should we understand and interpret these obstacles? Since previous studies regarding PBL implementation lack a systematic perspective to link PBL execution to the larger organizational context, the study in this book will explore PBL implementation within its organizational context, rather than as a teaching technique. Furthermore, this study argues that the challenges and obstacles organizations encounters are closely linked to the organizational members' interpretation of PBL implementation. The interpretations of PBL by members of the organization are reflected in their daily practice, rather than their conscious responses to the interviewer's question of "what are the challenges?" Given this consideration, this book attempts to explore how the organizational members understand and interpret the organizational change toward PBL.

In order to address these concerns, chapters 2 and 3 will conduct a brief review of the traditional wisdom regarding PBL and the theories of organizational change, specifically focusing on the strengths and the weaknesses of relevant theories.

Chapter 4 specifies our general research strategy, the data collection method, the data analysis process, and the research quality considerations. Case studies form the foundation of the research strategy of this book with an emphasis upon two universities which were in the process of transitioning from a Lecture-Based Learning approach to PBL and had differing organizational structures. The data collection draws from a variety of sources such as literature, interviews and observations. Additionally, this study, written in a narrative approach, intends to investigate both universities by exploring key areas such as why they made the decision to change, how the change progressed, what the challenges and obstacles of PBL implementation were, how the organizational members understood the change, and how their conceptions of PBL implementation interfered with the transition process. It is only through the demonstration of the detail of reality that we can understand these questions in more depth.

Chapters 5 and 6 report on the two universities and focus on the organizational change toward PBL which took place during the transition period. PBL was initiated by the top management at both universities, but with very different change strategies. For example, X University tends to impose a unified PBL approach throughout the university, while Z University encourages bottom innovations and explorations when implementing PBL.

Chapter 7 is concerned with the diverse differences in the understanding and interpretation of the universities in regard to PBL in general, and specifically the group work. Although both universities share commonalities such as leaning toward teaching rather than learning, emphasizing learning outcomes rather than processes, overemphasizing PBL's instrumental values, and implying a need of conceptual change of the organizational members, the differences in the interpretation of PBL are responsible for their different results (tensions/chaos).

Chapter 8 shifts our attention from the organizational members' interpretations of PBL to their understanding of motivation issues. It can be noted that the members of the organization were largely unmotvated to participate in PBL activities due to organizational factors such as the research oriented environment. The implication is that in order to facilitate a change toward PBL, it is not sufficient to change only the elements relevant to learning; rather, it is necessary to reflect on the role and function of the university.

Chapter 9 deals with the management of the transition process. The universities struggled between two contradictory approaches. On the one hand was the notion of the need to use a more "normative" approach to regulate staff behavior and secure the changes in the organization. On the other hand was the feeling that the staff needed to be given the freedom to implement PBL through innovations in their own classrooms and beyond. However, both approaches have their limitations: a strong normative approach can create major tension within the organization and too much individual exploration may lead to chaos. Therefore, the coordination and balance of these contradictory implementation approaches are always a challenging task.

Finally, chapter 10 will include a summary of and reflection on the major strengths and limitations of this study.

2

Basic Understandings of PBL

The first concern that occurs when we deal with the issue of PBL would be what it consists of since how we explore PBL is based upon our understanding of it. Therefore, a thorough review of the research documents relevant to PBL is needed so as to build a solid theoretical context for this study.

2.1 PBL: Definitions and Principles

This study begins by clarifying the definition of PBL in the domain of higher education. Many scholars have attempted to identify what PBL is. One of the earliest definitions uses five characteristics to identify PBL: complex and real-world situations, teamwork, students gaining new information through self-directed learning, teachers' role as facilitator and problems leading to clinical capacity development (Barrows & Tamblyn, 1980). Using this definition, PBL can be regarded as a teaching and learning method to achieve knowledge and capacity gains through the use of real-world situations and teamwork. It also proposes a way of shifting staff's identity from being instructors to facilitators. Later, Boud (1985) lays out eight characteristics of PBL:

- An acknowledgement of the base of experience of the learners;
- An emphasis on students taking responsibility for their own learning;
- A crossing of boundaries between disciplines;
- An intertwining of theory and practice;
- A focus on the process of knowledge acquisition rather than the products of such a process;
- A change in staff role from that of instructor to that of facilitator;
- A change in focus from staff assessment of learning outcomes to student self- and peer assessment;
- A focus on communication and interpersonal skills which help students understand that in order to pass on their knowledge, communication skills are necessary and go beyond their area of technical expertise.

Compared with the previous definition, this one is more concerned with the learning process and students' subjectivity. Within this definition, learning should highlight the process rather than the result and should be interdisciplinary and link theory and practice together. The fore-structure of the students' learning and the students' learning responsibility should be respected. It also details the learning outcomes in terms of communication skills. However, both definitions are flawed in the sense that they offer little indication of what distinguishes PBL approaches from the traditional education approach. To elaborate further on this, Walton and Matthews (1989) propose three dimensions to understand PBL: firstly, the requirement that curricula are organized around problems rather than disciplines, with an integrated curriculum and an emphasis on cognitive skills; secondly, the need for conditions that facilitate PBL such as small groups, tutorial instruction and active learning; thirdly, a focus on the actual outcomes in terms of skills and motivations, as well as the students' ability to become lifelong learners. In addition, Kolmos (2008) categorizes three major dimensions of PBL: a learning perspective based on dealing with problems, a content perspective in terms of interdisciplinary learning and social learning through group work. These attempts clarify the key aspects of PBL without losing the focus on its complexity and vitality.

By these definitions, problem centeredness lies at the center of PBL, which stipulates that the problem serves as the point of departure in the teaching and learning process, and then dominates it. In this sense, students are not simply asked to find the right answer or the solution to the problem; rather, it is through the exploration of the problem itself that students construct their knowledge and develop their skills. Furthermore, the problem also appears as a means to organize all learning material (across different subjects or disciplines). This stands in contrast to traditional educational approaches where learning content is organized around a particular subject or discipline. PBL, therefore, implies a new way of designing and developing curriculum. The curriculum is not just constructed based on isolated disciplines and subjects, but designed and integrated through problems and projects.

Learning within a PBL environment should be regarded as interdisciplinary, implying that students need to deal with the problem situations by using knowledge and methods from various fields. It is beneficial, indeed it is necessary, for students to break traditional disciplinary boundaries. It also implies the need for a new method of curriculum design in terms of strengthening the relationships between different disciplines, disciplinary integration, collaborative teaching, etc.

Finally, social learning requires students to work in groups. The benefit of this dimension is that students can develop their communication, collaboration, and problem-solving skills.

If we take a step further, we may note that these dimensions contribute jointly to a crucial educational principle: student centeredness, which transfers our focus from the teacher to the student when we consider the issue of learning. Traditional learning places the teacher in a dominant role throughout the entire learning process. Teachers design the learning content, control the knowledge delivery process, and behave as authorities. The drawbacks of the traditional learning paradigm have been widely criticized, citing as consequences such things as low attendance in the class, low motivation and reduced involvement in learning process for students. In contrast, PBL argues for student centeredness in the learning process which specifies that students have the ownership of the learning process. It is the student, instead of the teacher, who directs and controls the learning process. Student centeredness encourages the participation of students in learning because they feel more motivated. In some literature, student centeredness is also termed as self-directed learning, participate-directed learning (Barrows & Tamblyn, 1980; Donner & Bickley, 1993; De Graaff & Kolmos, 2003), students' taking responsibility for learning (Boud, 1985; Savery & Duffy, 1994), etc.

Although PBL is sometimes equivalent to a teaching and learning method, or an instructional model (e.g. Savery & Duffy, 1994), many studies suggest that the understanding of PBL should be shifted from merely a particular method of instructing to a general strategy. It is suggested that this approach includes a variety of different forms and types since PBL is highly dependent on the staff, the students, the discipline, the program, the organization, the culture, etc. To address this issue, we must bear in mind that there have been many attempts to develop PBL taxonomies (including different PBL curriculum models) explaining different PBL meanings, types and uses, etc. We can see these examples in, for example, Barrow (1986) PBL's taxonomy, Savin-Baden and Major's curriculum model (2004, p. 35–45), and Moesby's PBL approach (2004). Furthermore, to broaden the meaning of PBL, many studies (Margetson, 1991; Savin-Baden & Major, 2004) suggest that PBL should be seen as an educational philosophy. It is not just a matter of educational technique, nor is it a general term covering different sub forms. Rather, PBL represents a new way of conceptualizing notions of learning and educational institutions.

With regard to learning, as noted earlier, PBL is based on social constructivism (Vygotsky, 1962; Savin-Baden & Major, 2004), which means that knowledge is not conveyed from the external such as teachers' instruction;

rather, knowledge is acquired through the learners' self-construction which takes place in social settings. Through the thorough exploration of a particular problem situation in a PBL environment, learning is no longer based upon the memorization of a set of general propositions, but instead, can be viewed as dealing with specific cases in a concrete context which further facilitates the formulation of the learners' tacit knowledge (Flyvbjerg, 2006). Furthermore, although PBL is concerned with the improvement of learning outcomes (e.g. skill development), PBL still highlights the process of knowledge acquisition rather than the products of such a process (Boud, 1985).

Furthermore, within a PBL environment, we must shift our focus from teaching to learning, which can be further categorized as a paradigm change. Barr and Tagg (1995) note that the education field in the past century has undergone a paradigm shift from an instruction paradigm to a learning paradigm. The instruction paradigm focuses on educational institutions offering teaching, and its primary pedagogical concern is with making the instructional process more effective. The assessment, according to the instructional paradigm, should determine whether teachers have successfully delivered the prescribed teaching material to students.

However, the learning paradigm implies that the educational institutions' aim is to produce learning rather than teaching. Therefore, the university should be concerned with how to provide a supportive learning atmosphere, how to engage students in learning, and how to enhance learning outcomes. From the hermeneutical perspective, the paradigm shift from instruction paradigm to learning paradigm is necessary since education is not about what is delivered by teachers but what is interpreted by students (Gallagher, 1992). It is only in the interpretive process that genuine learning can occur. With regard to PBL, PBL is far more than a teaching technique since its emphasis on teaching places it well within an instruction paradigm. PBL strives to facilitate the necessary conditions for learning (e.g., small groups, teacher as facilitators) and foster students' learning outcomes (e.g., skill development, lifelong learning attitude). In this sense, PBL calls for a new way of reconstructing educational institutions, curriculum, administration, assessment and evaluation methods, resources, organizational culture, and all the members in the organization.

2.2 Problem-based Learning and Other Similar Terms

Problem-based learning is synonymous with project-based learning or project-organized learning. In practice, some universities such as Aalborg University employ a problem based learning approach combined with project-organized learning (Kolmos, Fink & Krogh, 2004). Boud (1985) argues that problem-based learning and project-based/organized learning refer to the same approach; meanwhile, Savin-Baden (2003, p. 18–19) insists that although both approaches have some commonalities, they differ from each other in the following aspects:

- Students using a project-based approach are required to produce an outcome in the form of a report or design. In PBL, the focus is on the process;
- In PBL, the tutor supervises rather than facilitates the project;
- In project-based learning students are required to produce a solution or strategy to solve a problem, whereas in PBL, solving the problem may be part of the process, but the focus is on problem management, not on defining a clear solution;
- Input from the tutor occurs in the form of some type of teaching during the project period in project-based learning. The focus in PBL is on students working out their own learning requirements. Some PBL programs use lectures as a means of supporting the students rather than directing their learning;
- Project-based learning often occurs towards the end of a degree program after certain knowledge has been covered and will be able to aid them in the project. PBL is normally used on the basis that students have already covered the required propositional knowledge; therefore, students decide what they need to learn;
- Project-based learning is often seen as a mechanism for bringing together several subjects areas under one all-encompassing activity at the end of the course. PBL departs from the premise that learning will occur across disciplinary boundaries, even at the beginning of the course;
- Project-based learning is more often seen as a teaching technique in a given area of the curriculum rather than an overall educational strategy such as problem based learning.

Another concept that is often synonymous with problem-based learning is problem-solving learning. Although they both involve dealing with problem situations, they differ from each other in many aspects (Savin-Baden, 2003, p. 2–3). For instance, within a PBL context, the problem is set as the starting point of the learning process and afterwards dominates the whole learning process. It is through dealing with the problem situation that the knowledge gradually appears and becomes an integral part of the student. PBL emphasizes the learning process rather than merely the learning outcome. On the contrary, the focal point of problem-solving learning is the learning outcome (solving the problem) rather than the learning process. The students acquire the knowledge beforehand and then they apply that knowledge to the problem situation. The problem-solving process essentially serves as a medium to examine the utility of the knowledge as well as the students' understanding of the knowledge. The major difference between problem-based learning and problem-solving learning can be noted as being that the knowledge acquisition occurs through dealing with the problem in PBL, whereas within the problem-solving learning context, knowledge is obtained before the problem is dealt with.

Within the medical field, PBL is often associated with case analysis or case study (Barrows, 1986). The educationalists in the medical domain tend to use the study of a medical case as a form of PBL. Barrow's (1986) taxonomy of PBL approaches addresses the different uses of a medical case within a PBL context. The first PBL approach is lectured-based cases where the information is presented first and afterwards case material is used to demonstrate it. The second approach uses case-based lectures to present students with cases covering the necessary material. The third approach is the case method in which students are assigned a case to research and prepare for discussion. The fourth approach is modified case-based where students are presented with some information and asked to decide on the actions and decisions they would make. The fifth approach is problem-based and has students meet with a client for a free inquiry. Finally, the sixth approach is a closed-loop problem-based method, which is an extension of the problem-based method, where students are asked consider the resources they used in the problem-solving process in order to evaluate how they may have reasoned through the problem more effectively. All of these approaches have PBL components. Yin (1994) highlights the significant distinction between case studies as a teaching method and case studies as a research strategy. For teaching purposes, a case is used to build a framework for discussion or debate among students, and it does not need to contain a complete or accurate account of an event. The criteria for the selection and development of a teaching case are far different from that of a research case. However, case studies in both teaching and research fields can be regarded as a process of learning and knowledge construction. Whether they are aimed at classroom discussions or for the purposes of research, they represent a particular way of learning: learners construct knowledge through an intensive inquiry of a particular, often real-life event. Using case studies as a research strategy will be further elaborated upon in chapter 4.

2.3 Theoretical Foundations for PBL

Theoretically, PBL can find its roots in a variety of psychological theories and pedagogical works such as constructivism, cognitivism, experiential learning, and humanism, among others (Savin-Baden & Major, 2004). Each theory proposes a particular way of dealing with education. If we only concern ourselves with the psychological way of dealing with education, the fact that the education has a linkage with the learners' life experience would probably be neglected. Education, however, according to Dewey (1916, 1938), is for human growth which is inclusive of far more than the change or the development of learners' psychological dimension. Growth for behaviorists only accounts for improvements in the learners' behavior regardless of their internal process. Cognitive theorists note the function of learners' cognitive structure, but somewhat reduce the growth of the learner to merely the development of the cognitive structure. Social constructivism theorists recognize the role of social communication in the growth of one's cognitive disposition (Vygotsky, 1962). Humanists have the tendency to comply education with human's natural needs and impulses such as emotions. Maslow's *hierarchy of human needs theory* (Jaffee, 2008) hypothesizes that the internal law of human development is not always a compromise between the so called natural impulse and law and the reality of the social setting. Humans cannot grow independently of their social environment, thus calling for a new way of looking at education.

This section primarily focuses on the philosophical foundations of PBL and specific attention is paid to PBL John Dewey's theory of experience (1916, 1938), since PBL manifests the basic principles of what he terms "a good experience."

Concerning education, Dewey proposed an innovative approach to understanding education by conceiving it as an experience. The use of the term *experience* suggests the need to unify the subjective matter and the external environment. Education intertwines with the learners and does not reside outside of oneself. Since education causes us to experience something, it is necessary to separate the good experiences from the bad experiences, which Dewey terms educative and mis-educative, respectively.

According to Dewey (1938), a good experience is characterized by two principles: continuity and interaction. The importance of continuity in human intellectual development has been captured by many theorists but is usually confined to a narrowly psychological sense. For instance, Piaget (1953) notices the continuity of the development of human's cognitive structure, or in his term, *schema*, as shown in his cognitive developmental theory. Ausubel and Robinson (1969) contend that meaningful learning is more likely to occur when there is a link between the learning material and the learner's prior knowledge. Therefore, it is essential for the educator to offer appropriate information to bridge the learner's prior knowledge with the new learning material. While Ausubel and Robinson (1969) narrow the principle of continuity to its cognitive meaning by highlight the importance of continuity of the learners' cognitive process and contending that knowledge construction should be concerned with the learners' existing cognitive structure, they, meanwhile, expel the learners' life experience from the learning process.

The notion of continuity in Dewey's work refers to the fact that "every experience both takes up something from those which have gone before and modifies in some way the quality of those which come after" (1938: p.140). Therefore, current experience is a consequence of or largely conditioned by a previous one, and at the same time it may direct future experience by affecting one's attitude and objective conditions. Good experiences involve the formulation of habits and attitudes which constantly facilitate the unfolding of continuing experiences. In this sense, traditional education deviates from the principle of continuity as it leads to the production of a great many negative habits and attitudes (rote learning, low motivation, etc.), which may restrain the growth of future experiences.

PBL can be regarded as an attempt to maintain continuity of the student's educational experience. PBL attempts to bridge the gap between theory and knowledge (Boud, 1985), a concept which is distant in the traditional educational context, where the knowledge that is imparted to the students is separate from their life experience. PBL sets the exploration of the problem scenario or the real-life situation at the center of the learning process and through the problem-solving process, students are able to find the practical meaning of the knowledge they are using. In this sense, the knowledge is no longer independent of the students' lives, but an integral part of their experience. Furthermore, PBL is widely evidenced as being efficient in providing students with a set of general skills in terms of communication, critical thinking, and problem solving, which facilitate the students' employability and professional life. In addition to this, PBL can promote the students' learning motivation and cultivate a lifelong learning attitude which can direct their future in a positive manner, or in Dewey's terms, it is conducive to students' growth.

The principle of interaction highlights the importance of one's engagement with the external environment which is "whatever conditions interact with personal needs, desires, purposes, and capacities to create the experience which is had" (Dewey, 1938, p.42). The environment can refer to the people, the subject, the experiment, the activity, or whatever object a learner may interact with. Dewey thus rejects both traditional education which, he noted, had "emphasized the external conditions that enter into the control of the experiences, but that it paid little attention to the internal factors which also decide what kind of experience it had" (p.39) and progressive education which overemphasize one's needs and impulses. With regard to PBL, the object with which the students primarily interact with refers to other human beings including students and teachers. The learning occurs better when there is constant interaction between the learner, other learners and the teacher. In order to understand this concept better, we will continue to explore the concept of social constructivism which is closely linked to Dewey's notion of interaction and widely regarded as a key theoretical foundation of PBL.

2.4 PBL, Social Constructivism and Dialogical Learning

PBL is theoretically based upon social constructivism (Savin-Baden & Major, 2004). From this viewpoint, the nature of the knowledge is not independent of man's mind, but is a product of the learners' social construction. In this sense, knowledge is no longer purely objective and static but changeable and dependent upon the context. Moreover, it points to a new approach to obtaining knowledge since the traditional epistemology views knowledge as being obtained from simply reasoning, as rationalism suggests, or interacting with the non-human world, as the positivists suggest, that knowledge needs to be validated by scientific evidence (such as experimentation, observation, etc.). Social constructivism, however, suggests that social interaction among people can serve as a powerful means to formulate knowledge. As Habermas (1984) notes, interaction, or the communication between different human beings is another crucial dimension in formulating knowledge and rationality. The concept of interaction is crucial to understand social constructivism as well as the social learning dimension of PBL.

In cognitive constructivism, such as Piaget's theory (1953), it is suggested that learning is not the transmission of knowledge to the learners since it is not ready for the student to understand and use. Learning is, rather, the personal continuous reconstruction of one's cognitive structure through accommodation and assimilation, which can also be termed individual constructivism.

Unlike this theory, social constructivism emphasizes the importance of social interaction for learning (Powell & Kalina, 2009). Social interaction is widely recognized as an essential factor in promoting student learning. Vygotsky (1962) notes that students tend to learn better when they can obtain assistance from their teachers or other students. The implication is that the interaction with others can be a powerful means of enhancing students' learning. Therefore, cooperative learning is an integral part of achieving a better understanding of the knowledge. The relevance of social constructivism to PBL is widely recognized in the form of social learning, which is generally taken as one of PBL's major traits (Walton & Matthews, 1989; De Graaff & Kolmos, 2003; Kolmos, 2008). The term social learning should not be understood in the psychological sense that the learner gains knowledge of action through observing the behavior of others (Bandura, 1986). Social learning within the domain of PBL generally refers to a learning approach where students work on a common problem scenario in the form of a group usually consisting of four to six students (Walton & Matthews, 1989; De Graaff & Kolmos, 2003; Kolmos, 2008). As students in a PBL group work on a common problem, they present, question, and challenge each other's viewpoints. They are also expected to divide tasks among each other and cooperate towards the same goal. Through group work they learn to manage a team while learning to respect, listen to and understand the members of their group. It is believed that through group work, students are able to develop their communicative skills, teamwork skills, group management skills, etc. Furthermore, a social learning approach can contribute far more than the simple exchange of ideas and thoughts, or the cultivation of communication skills, but also serves as a means of promoting students' mutual understanding. Since each participant brings his or her own context of understanding into the group, a social learning approach enhances the fusion of the horizon (Gadamer, 1995) among the group members.

Another important factor, which is often somewhat missing in PBL literature, is the importance of the tutor, or the facilitator. Here, we are not suggesting that there is a lack of literature with regard to the tutor in general since there are abundant studies discussing the role of the tutor, how the tutor should facilitate the student's learning and how to equip the tutor with adequate facilitating skills (Savin-Baden, 2003; Connolly & Silen, 2011). Instead, we contend that the tutor is seldom discussed as an element of social learning. The common notion is that the social learning primarily refers to the student group with the implication that the tutor resides somewhere outside the PBL group. This book discusses the tutor not as a tool *for* social learning, but as an integral part *in* social learning.

If we come back to Dewey's term interaction, the external environment that the student interacts with is, by nature, inclusive of the tutor. Furthermore, when we are dealing with a PBL approach where the tutor should be actively present, the tutor in the social learning environment cannot be neglected. Therefore, the tutor should be considered an insider, an integral component of the social learning in the group. The relationship between the teacher and the student in a traditional educational approach is characterized by a hierarchical structure where the teacher represents the knowledge and the authority and is seldom challenged by students who have a lower status. This teacher-student relationship is described by Freire (1972) as oppressing and detrimental to student and teacher interaction. However, when the teacher appears as the facilitator, he or she is no longer the authority or the oppressive force that hinders effective interaction in the classroom, but serves as a participant in the learning process. Therefore, the social learning of PBL does not only refer to the interactivity among the students within the group, but also includes the tutor in the PBL group.

Another important term that is often associated with social learning is dialogical learning, which, though it embraces the relationship between the students, is predominantly concerned with the relationship between students and teachers. The research field is replete with studies dealing with the dialogical learning (Shor & Freire, 1987). There are also a great many recommendations of how to design a better dialogical learning environment such as guaranteeing students' equal participation, etc. Dialogue or communication primarily occurs, not as a method for human interaction, but as an ontological human concern. As Freire (1972) states, the human is by nature dialogic and communication is crucial for a human being's life since it is through the dialogue with others that we formulate our identity. He stated,

> "Dialogue must be understood as something taking part in the very historical nature of human beings. It is part of our historical progress in becoming human beings. . .(through dialogue) humans become more and more critically communicative human beings. Dialogue is a moment where humans meet to reflect on their reality as they make and remake it."
>
> (Shor & Freire, 1987, p.13)

Furthermore, dialogue should not simply be regarded as a technique to achieve some cognitive outcomes, but as a means to transform social relationships within the classroom as well as in the societal context (Freire, 1987). In this sense, the notion of interaction is crucial since it calls for a new way of looking at the relationship between authority (i.e., the teacher) and the student. Freire (1972) argues against the oppression of the teacher upon the student within the traditional educational context. In order to create a democratic educational atmosphere, the traditional teacher-student relationship should be challenged and restructured through the use of a dialogical learning approach (Shor & Freire, 1987). In addition, the objects with which learners dialogue are not confined to just teachers. Gadamer (1992) extends the object of dialogue from the human being to the text and contends that the human's understanding of behavior significantly relies upon the communicative process between one's self and the object that he/she is attempting to understand. In this sense, the relationship between the students and all kinds of knowledge authorities of all forms, namely teacher, textbook, or senior student, should be restructured and treated in a dialogical manner.

To summarize briefly, social learning involves far more than the interaction between students since we take the teacher/tutor as an integral element of the social learning of PBL. The value of social learning lies, not only in the fact that learning can be improved by an environment conducive to a social interaction (Vygotsky,1962), but also that social learning, or what is termed as dialogical learning (Freire, 1972, 1978) can serve as a means of transforming the relationship between the learner and the authority (including the teacher, the textbook, and so on). Finally, PBL's social learning must take into consideration the interaction and communication between the students and the tutor/teacher.

2.5 PBL Models

PBL is seen by many researchers as far more than a single instructional approach; it is an educational philosophy that covers many different forms and variations in practice (Barrows, 1986; Savin-Baden, 2003; Moesby, 2004; Savin-Baden & Major, 2004). The traditional approach to designing curriculums (see Figure 2.1) is mainly based on a single discipline. The curriculum system in a given semester consists of several independent courses/subjects delivered by different tutors from different disciplines. The learning focus is on the knowledge, theory, and the fundamental content of

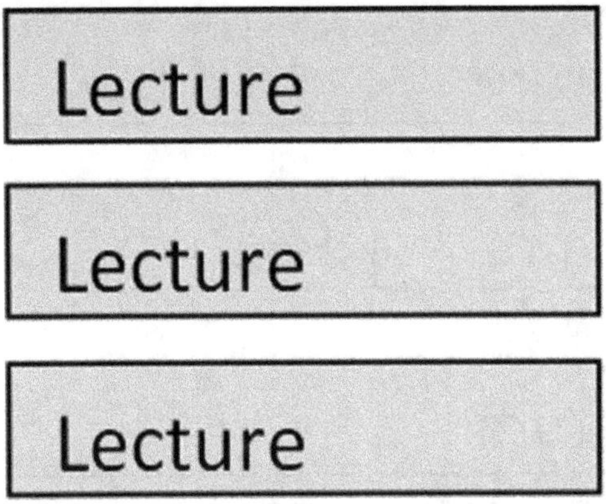

Figure 2.1 Traditional Curriculum with Independent Lectures

the curriculum. The assessment method consists of testing how much students are able to memorize from the course.

The second approach to curriculum design (see Figure 2.2) is comprised of several independent courses. A course is composed of a lecture unit which is followed by a PBL unit, aiming at applying knowledge learned in the lecture to a problem situation. In this model, knowledge retention is still seen as the priority of all educational objectives. PBL components only serve as a means of testing the quality of knowledge retention or to improve students' understanding of the learning material of the lecture. It emphasizes the importance of the basic theory and ignores practice. An additional problem of this model is that there is a lack of coordination between different subjects since each course has its own particular learning contents and learning objectives. Another possible danger is that it can lead to an overwhelming burden for the students when several PBL units coincide.

In order to strengthen the relationship between different courses, build a strong connection between theory and practice and avoid the possible danger of creating too much academic burden for students, several approaches have been developed. One possible way of avoiding some of the above mentioned pitfalls is to have several courses establish a common PBL unit (see Figure 2.3). The courses are still delivered separately by different staff members, similar to traditional education, but they are coordinated by a common PBL unit in terms of a project or a case. The relationship between different courses is manifested through PBL processes, as students have to deal with a problem situation by using the learning contents from several courses. However, although PBL becomes an independent component in this model, the learning focus is still

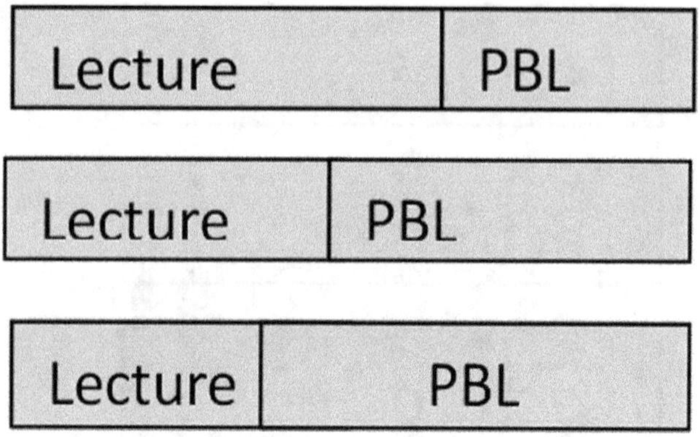

Figure 2.2 Segmented PBL curriculum

on knowledge retention. The PBL aspects in this model are no more than a field for knowledge application.

The last PBL curriculum model shifts the focus of curriculum design from knowledge to PBL (see Figure 2.4) PBL. In this model, curriculum development starts from the design of the problem in PBL. After the problem has been identified, courses are meticulously chosen to facilitate the learning objective. In this sense, PBL can be regarded as an educational philosophy since the problem guides the whole curriculum development process.

2.6 Aalborg University PBL Model

Despite the different PBL curriculum models, we will particularly focus on the Aalborg University PBL Model in order to elaborate on how PBL works in practice, as it can serve as a "benchmark" or "yardstick", or an ideal type[1] of PBL. The reason for choosing the AAU PBL Model is not because it is the only correct way of implementing PBL. Rather, since Aalborg University has nearly four decades of experience practicing PBL throughout the university, its PBL model, to some extent, exemplifies essential PBL principles and can serve as a starting point for understanding and comprehending what PBL means and how PBL works in a real-educational context.

[1]The notion of ideal type is developed by Max Weber (1864–1920) in his analysis of capitalism society. An ideal type is primarily a mental construct that allows one to highlight the crucial elements of a complex social phenomenon. In this study, AAU's PBL Model can be regarded as an exemplar of an ideal type of PBL approaches.

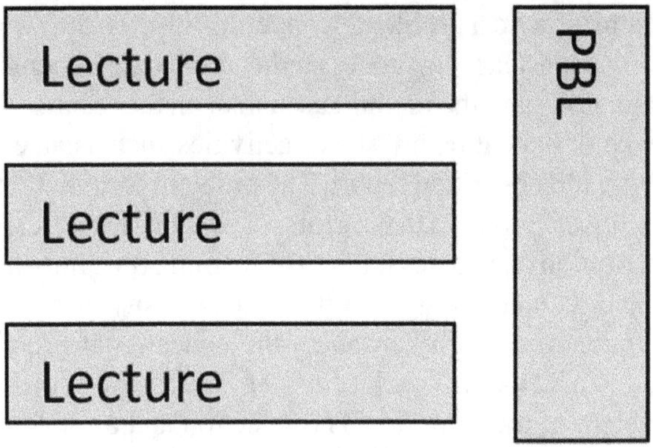

Figure 2.3 PBL Curriculum with Common PBL Components

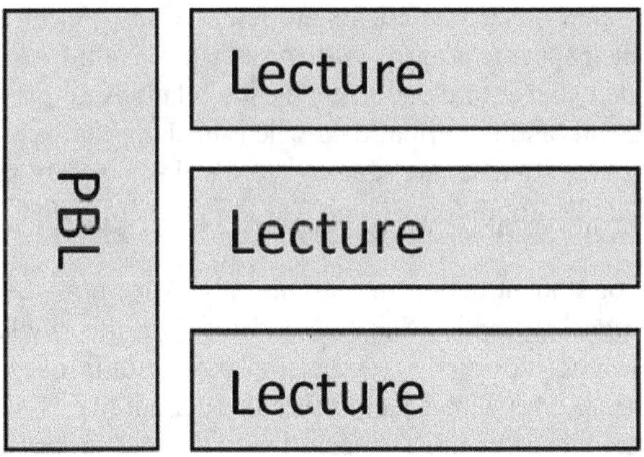

Figure 2.4 PBL Model Where PBL Components Direct the Curriculum

2.6.1 AAU PBL Vision

PBL principles are explicitly articulated in the educational vision of Aalborg University (Barge, 2010). In the description of its vision, Aalborg University specifies six central principles as its basic features: problem orientation, project organization, integration of theory and practice, participant direction, team-based approach and collaboration and feedback.

More accurately, the Aalborg University PBL Model (AAU PBL) can be seen as a combination of problem-based learning and project-organized learning. The problem stands as the basis or the starting point of the whole learning process. The problem can be presented in a number of ways. It can be a theoretical problem, practical problem, wondering, discrepancy, etc. The primacy of the problem in the learning process indicates that PBL focuses more on the problem formulation[2] and the method used than on the results. The project could be a sequence of educational tasks or activities such as an engineering product or a well-organized written report. The problem formulation serves as a means of addressing the problem that a group of students propose. At Aalborg University, the institution distributes half of the learning time into the project.[3] Problem and project centeredness at Aalborg University serve as a guide for curriculum arrangements. For instance, the educational problem at the university should have a "theme" each semester. The theme is the main topic or the overall problem of the semester. The selection and arrangement of the curriculum, whether it is general curriculum or a project-related curriculum, should be relevant to the semester theme in some way.

To link theory and practice, students are required to connect theoretical knowledge with their specific projects with the assistance of their supervisors. For instance, students have to clearly specify the relationship between their theoretical framework and empirical data when finishing the written report. Through this practice, students are able to see the link between the knowledge gained in the classroom knowledge and the problem found in the real world.

The primacy of student action in learning, also known as the principle of student centeredness, implies that the students own the whole learning process. External authorities such as teachers, textbooks, or institutions do not guide students; rather, they direct themselves throughout the whole learning process, though there is an exam. The team-based approach refers to a social approach which requires students to work in groups or teams. Students benefit from group work in a number of ways: students can construct knowledge through interaction with other group members, which is a basic assumption of constructivist theory; students gain a certain set of skills which are more likely to be developed through group work such as communication and collaborative skills; and students can discipline each other, thus encouraging each other's motivation and engagement in the learning process.

Collaboration and feedback, according to the vision statement, refer to the outcome of the PBL model. Through PBL, a specific set of skills should be developed: collaboration, feedback and reflection.

---

[2]A problem formulation is a problem statement.

[3]A project at Aalborg University can mean a task, a product, or a written report; students, in the form of a group, must contribute half of their semester to the project work and hand in a final product for assessment at the end of the semester.

2.6.2 AAU PBL Model

PBL is explicitly represented by its curriculum system. The curriculum includes a course unit and a project unit, and each part accounts for half of the curriculum. The course unit can be further divided into study-related courses and project-related courses, each of which is approximately 7.5 ECTS per semester. The study-related courses aim at offering students fundamental knowledge and the project-related courses are supposed to be directly linked to project work. Normally, students attend both types of courses during the first half of the semester, and work on the project in small groups during the second half of the semester. Students are required to apply what they learn in the courses to a project, which is a way of showing them the connection between theoretical knowledge and empirical work.

3

The University as an Organization

PBL is always implemented into a particular organizational setting, and therefore, our understanding of PBL implementation is highly dependent upon our knowledge of the organization's structure. With an understanding of the structure we can then look at the transition process to PBL to understand the complexity and dynamics of the process of change within the organization. The necessity of understanding organizational structures and processes directs us to organizational theories. Currently, the research field is replete with assumptions, perspectives and approaches regarding how to perceive and analyze organizational phenomena and functions (Jaffee, 2008; Van de Ven and Poole, 1995; Dalin, 2004). Drawing on the wealth of research into the complexity and dynamics of the process of organizational change to PBL, this chapter will focus on some influential and representative organization theories and approaches on a general level, while paying particular attention to each theory's strengths and weaknesses in depicting and analyzing the change process. By knowing how current organization theories describe the change process as well as their strengths and limitations, we expect to develop an approach that can explain and interpret the educational transition as a dynamic process instead of simply describing the static factors related to it.

The organization research field has abundant definitions of the organization, and some of them are extremely complicated. For example, Hall put forward "An organization is a collectivity with relatively identifiable boundary, a normative order (rules), ranks of authority (hierarchy), communication system, and membership coordinating systems (procedures). This collectivity exists, on a relatively continuous basis in an environment, and engages in activities that are usually related to a set of goals; the activities have outcomes for organizational members, the organization itself, and for society" (Hall, 1999:30). The complexity of such a definition is confusing; and a thorough exploration of a number of definitions of organizations is overwhelming. Dewey (1916) pointed out that "Science has been defined in terms of method of inquiry and testing", rather than "organized or systematized knowledge" (p. 190). Thus, it is the methodological consideration of, or the approaches to the focal research phenomenon, that defines the final definition of the research object. Therefore, rather than offering a single definition of organization as the point of departure, it is intriguing to consider how one might reach an understanding of the organization without a definitive definition. This section will, therefore, outline the conventional wisdom of the approaches and the perspectives of the existing organization theories.

3.1 Different Organization Profiles

3.1.1 Scott's Five Elements for the Organization

Scott (1987) referred to five fundamental elements of the organization: social structure, participants, goals, technology and environment. Each element points out a possible approach accessible to the organization. The social structure refers to the patterned forms of the activities, relationships and interactions formally created in the organization. These structures are the basic building blocks of the organization, distinguishing the organization from other temporary collections of individuals. The participants are the actual bearers of all organizational functions, and the organization cannot survive without the existence of the participants. Goals are the ends the organization seeks to obtain, and they significantly steer organizational activities and performance. Technology includes all kinds of means, techniques and procedures that process the organizational raw material and transforms the organization's input into output. The environment of the organization significantly supports and confines its performance and fate as it goes through economic, political, social, legal, cultural and organizational systems. Each of these elements, by and large, points to a central feature of the organization and indicates a possible perspective of the organizational analysis. We can see that many organization analysis approaches can trace their prototype here. In the subsequent section, we discuss how different organization theories are actually directed to one or several of the five elements above. Scott's (1987) framework is based on the principle of dissecting the organization structure, a prevailing method in current organization research. This method can be used to conduct a structural analysis of the organization, dissecting the organization into different elements, analyzing them in detail and integrating them into a whole.

Yet, this framework is flawed by its weakness to describe the organizational process.

3.1.2 Organization as Bureaucracy

This perspective is largely concerned with the formal structure of the organization in terms of hierarchy, authority, distribution of labor and tasks, etc. One of the most classical and influential attempts depicts the traditional organization as the rational-bureaucratic model with the intention of attaining maximum effectiveness. Max Weber outlines six essential elements of bureaucracy: "(1) clearly defined division of labor and authority, (2) hierarchical structure of offices, (3) written guidelines prescribing performance criteria, (4) recruitment to offices based on specialization and expertise, (5) office holding as a career or vocation, and (6) duties and authority attached to positions, not persons"(Jaffee, 2008:90). Currently, this model is the predominant organizational model in governments, factories, firms and hospitals. In the educational field, many researchers borrow this model to elaborate upon the structure of the educational institutions. Stroup (1966) generalizes some of the characteristics of universities and colleges as following: (1) Competence is the criterion used for appointment; (2) Officials are appointed without election; (3) Rank is recognized and respected; (4) Careers are generally exclusive and do not include work done in other places; (5) One's lifestyle is centered around the organization; (6) Security is present in a tenure system; (7) Personal and organizational properties are separated. This bureaucratic model is organized around rational principles such as formal rules, procedures and a well-defined hierarchy. The hierarchical arrangement can serve as the most effective method to distribute people and tasks at complementary layers (Jaques 1990). An integral part of the bureaucratic organization is the scientific management principles developed by Frederick Taylor, which are still retained in modern society and assure the effectiveness and productivity of many organizations. The principle of formal structure, hierarchy, division of labor, distribution of tasks and standardized processes still plays a vital role in current organizations. A typical example of this model is the MacDonald's model.

However, this approach has some drawbacks. Much attention has been paid to formal authority instead of informal power generated from non-legitimate threats, mass movements, expertise and individual personality; thus, this model lacks the interpretation of the process of the organization structure, and will likely fail to explain how organizations and policies change

(Baldridge, Curtis, Ecker and Riley, 1978, p.30–31). This perception views the organizational change as a rational and linear process guided by prescribed objectives and procedures in order to attain the desired outcomes and reach the expected ends. However, it overlooks the complexity and dynamics of organizational change. The change process cannot be fully controlled and predicted. Change can be planned, but how the change process will unfold is far beyond rational control. The uncertainty of the external environment and internal environment complicates the rational control of the organizational change.

3.1.3 The Organization as a Human Organization

The introduction of the human factor into the organizational analysis is another powerful and influential approach which is able to shape how the organization's structure and the management method are implemented in practice. Furthermore, adding the human factor also transforms our perception of the organization process. This approach reveals the close relationship between human factors and organizational performance. The Hawthorn experiment (Jaffee, 2008) discovered the human dimension of organization, suggesting that humans perform better when they receive special attention. The need for attention should be taken into account when considering how to promote effective organizational change. Chester Barnard (1938) noted the tension between the individual and the organization due to the incongruence of individual interests with organizational objectives. The organization, therefore, has to make efforts to reconcile the different interests of its members with the aims of the organization. Human relations and human resource theories embrace the reality that humans have both social and personal needs, as proposed by Maslow's human needs hierarchy (Jaffee, 2008). It also stimulated the emergence of leadership theories, which have sought to bring together leadership style and organizational context.

The ideal type of human organization in higher education is the collegial model, based on the conception of a community of scholars managing its own affairs (Wolotkiewicz, 1980, p.174). As Birnbaum stated, "An emphasis on consensus, shared power, common commitments and aspirations, and leadership that emphasizes consultation and collective responsibilities are clearly important factors at (a collegial university). It is a community in which differences in status are deemphasized and people interact as equals, making it possible to consider the college as a community of colleagues-in other words, as a collegium" (Birnbaum, 1988, p.86–87). Since the staff members have the ownership of the organization, and organizational decisions are based on consensus of the staff members instead of managers, they have a deep commitment to organizational affairs.

Compared with the bureaucratic model and the scientific management notions of organizational change bringing advancement towards the organization's intended ends, human organization considers the human factor in the organization to be crucial to any organizational process. Without concern for human dimensions related to change and the potential tensions change can bring to an organization, organizational change is much more likely to encounter difficulties. An appropriate approach to organizational change would thus be concerned with its organizational members' needs, facilitating cooperation and collaboration and reconciling the objectives of its members with that of the organization.

A critical perspective presumes the organization to be political and to consist of different interest groups. This approach disputes model of the coherent, consistent, and unified organizational profile as described in the bureaucratic approach. Regarding higher education, Baldridge (1971) takes into account conflicts that occur between different groups of members of a college and thus confirms the political viewpoints about the college by drawing on three sources: conflict theory, community power studies and interest groups. Here, the single, clear goal in the bureaucratic organization is replaced by multiple objectives subject to fierce conflict. From the political perspective, the college can be described as follows: "there is a complex social structure that generates multiple pressures, there are many sources and forms of power and pressure that impinge on decision makers, there is a legislative stage that translates these pressures into policy, and there is a policy execution phase that generates feedback and potential new conflicts" (Baldridge, Curtis, Ecker & Riley, 1978, p.41). Some characteristics of the political model are: (1) high specialization and diversity, (2) the tendency of different interest groups to negotiate with each other so as to realize their goals and aims, (3) an unstable organization with intense disagreements and conflicts, (4) highly fragmented and decentralized power structure, (5) extreme influence from external forces and pressures. As a result, some factors should be focused on when policy making is ongoing, such as the mechanics of goal setting and the conflicts over values, change processes and the adaptation of the organization to its changing environment, conflict and conflict resolution, the role of interest groups, and legislative and decision-making phases (Baldridge et al., 1978, p.41). Organizational change, from this point of view, can be regarded as the replacement of the objectives among various interest groups in the organization.

3.1.4 The Organization as an Open System

The importance of the relationship between the organization and its external environment results in an open-system approach, containing the hypothesis that the organization is situated within an environment to which it must comply. This environment is largely seen as conditioning the organization's performance and development. The notion of an open system indicates that the organization is not a closed system; rather, the organization exchanges material, information and humans with its surrounding environment. Contingency theorists argue that the structure of an organization is largely conditioned by its outer environment. The organization has to alter its objectives, its structure and its actions in accordance with its environment when external situations change. Disputing the environmental determinism tendency of contingency theories, resource dependence theorists, advances proactive strategies that can be employed to deal with environmental constraints (Jaffee, 2008) while simultaneously acknowledging the organization's reliance upon its external environment in terms of resources. In general terms, the contingency theory portrays an organization profile as passively adapting to the environment, while the resource independent theory assigns the organization a positive role in competing for resources with their counterparts. New institutionalism theorists place an organization into the institutional framework, significantly confining the organizational structure and its practices. In this sense, the organization, on the one hand, has to struggle for maximum effectiveness, while on the other hand, has to conform to its external institution in terms of values, norms and culture so as to gain legitimacy and acceptance. The institution is described as an established order comprising rule-bound and standardized behavior (Selznick, 1996). The new institutionalism theorists thus contend that in addition to resources, the intention of being accepted and admitted by the institutional environment is especially crucial to the development of the organization.

Relating the open-system theory to organizational change, it is worth noting that organizational change can be viewed as a result of its response to the environment. No matter what external mechanism is present (resource, institution, political situation, etc.), the organization has to function and behave in accordance with its outer environment, either in an adaptive manner or in a proactive manner. The forces from the outside are the determinants of how the organization transforms. However, the open-system theory does not help us to understand the internal processes that take place when an organization makes a change, nor does it take into consideration the human factors at play in the change process.

3.1.5 The Organization as Anarchy

The notion of anarchy is concerned with the uncertainty and ambiguity of the organizational phenomenon and process, which sets it apart from other approaches that assume rationality, certainty and controllability of the organization. The organization is viewed as an organized anarchy, characterized by "problematic preferences, unclear technology and fluid participation" (Cohen, March & Olsen, 1972). The ambiguity of organizational goals, power, experience and success (Cohen & March, 1974, p.195–203), call into question the widely-accepted ideas that "organization leaders play a critical role in institutional processes, that institutions have goals, that individuals can specify their preferences, that chains of cause and effect lead individuals and organizations to take certain actions in order to effectuate outcomes they consider desirable, that problems are solved by decisions, and that decision making is a primary occupation of organizational participants" (Birnbaum, 1988, p.154). Relating this to higher education, Birnbaum lays out several main characteristics of the higher education institutes: the ambiguity of objectives in the college, the uncertainty of technology, and the change of participants in the decision-making process (Birnbaum, 1988, p.155–156). Uncertainty and ambiguity is further reinforced by the fact that large and complicated organizations, especially higher education institutions, consist of subsystems that are "loosely coupled" to each other. These subsystems are attached to each other and are somewhat interdependent, while they still retain their individuality, identity and independence. Under this organizational profile, attempting to make predictions of organizational behaviors becomes extremely challenging, even impossible.

Decision making in organized anarchy is often described as the garbage can model, and has four streams: (1) a stream of choices characterized by an entry time and a decision structure, (2) a stream of problems characterized by an entry time, an energy requirement and an access structure, (3) a rate of flow of solutions, and (4) a stream of energy from participants (Cohen et al., 1972, p.3). An anarchic institution randomly experiences a flow of the four streams. A problem is solved when one of a set of solutions meets the right participant who is available. Decisions can be viewed as random combinations of choices, problems, solutions and participant's energy. The time pattern of the problem's choices, the solutions and decision makers, the allocation of energy from participants and the linkages between those variables can influence the garbage can decision-making process (Cohen et al., 1972, p.4). Decision making in this model is characterized by (1) indifference by most people toward most issues on most occasions, (2) inertia of the entire system, (3) a great possibility that any decision will becomes a garbage can for almost any problem, (4) vulnerability of choice processing to overload, (5)weak information base (Cohen & March,1974, p. 206–207).

The anarchy model argues against the coherence and consistency of the traditional organization profile. Though it does not deny the rationality of individual human beings, the uncertainty and ambiguity of the organizational system (objectives, technologies, participants, time, problems, solutions, etc.) largely blur the organizational process. Therefore, changing motives, participants in charge, and the moment of the decision making, make the path of change completely unpredictable. This model pays particular attention to the complexity and uncertainty of the structures and processes of an organization. However, it contributes little to directing and guiding the process of change because it focuses more on description rather than on providing solutions.

3.2 The Process of Organizational Change

The aforementioned approaches view organizations as rather static, and therefore neglect the organizational process itself. This section attempts to address the organizational process. There are many explanations regarding organizational change as a sequential process in a number of domains, especially in business and management studies. Fullan (2001, p. 50) describes the change process as three successive stages: initiation, implementation and institutionalization. Initiation refers to the process of planning and making decisions, implementation involves putting the plan into practice, and the institutionalization phase is when the results of the change become an integral part of the organizational routine. He also specifies the crucial factors that determine the change's fate in each stage. More pragmatically, Kotter (1995) and Cohen (2005) outline eight phases for implementing change in an organization: creating a sense of urgency, forming a powerful guiding coalition, creating a vision, communicating the vision, empowering others to act on the vision, planning for and creating short-term wins, consolidating improvements and producing still more change and institutionalizing new approaches. Triggering and maintaining momentum among all organizational members is the key factor in the success of an organizational change. Relating the phase model of organizational change to PBL implementation at higher education institutions, Moesby (2004) proposes that the introduction of PBL at a university can be organized into four phases: Investigation (pre-action activities), adoption (vision formulation, defining success criteria, communicating the results), implementation (staff development program, evaluation program) and institutionalization. Though each stage possesses a particular set of objectives and activities, these attempts center around one major issue of how to initiate, retain and sustain the momentum for change. The ultimate resource for this momentum neither comes from the senior managers nor external pressure, but rather, from the ground-level organizational members. Without their enthusiasm and commitment to the organization, no change plans would move forward. Therefore, efforts to intrigue and maintain people's engagement in the change process should be made continuously. These models are "instrumental" tools for directing and guiding the implementation of an organizational change plan, but they provide little detail for us to appreciate what really happens during organizational change.

3.3 Levels of Organizational Change

Kumar and Thibodeaux (1990) identify three levels of organizational change. The first level targets a limited number of aspects of a single subsystem without causing any major change to the targeted subsystem or to the organization as a whole. The current structure and the value of the organization remain unchanged. The second level of change involves a major change to a particular subsystem of an organization rather than inducing organization-wide interference. The third level of change involves diagnosing and assessing the current organizational structure, as well as considering its configuration, patterns and value system. It is likely that the third level will lead to large-scale adjustments and modifications such as the elimination of some subsystems of the organization, or at the very least, significant alterations in the subsystem. Differentiating these levels of change is not simply to identify the change scale, but also has implications for action. The change planner should be aware of which level of change they will work at and carefully choose the appropriate strategy to facilitate change.

Moesby (2004) argues that the implementation of PBL in higher education institutions can be identified at three distinct levels. At the individual level, change happens when a limited number of teachers partially adopt PBL elements in the spirit of innovation and experimentation within the traditional framework. At this level, the traditional structure, beliefs and values are not questioned or challenged. The change at the next level, the system level, can be regarded as a transitional phase where the organization embarks on partially considering and coordinating PBL elements in a systematic manner such as transforming its assessments methods. Nevertheless, in order to involve the whole organization, the institutional level should be taken into consideration. This includes encouraging student participation at the management level, supporting interdisciplinary learning, making changes to evaluation and assessment methods, and organizing a cultural transforming, etc. It is the all-encompassing nature of the organizational change that distinguishes the institutional level change from the other two levels.

The typologies of the change levels seem to differ from that of the change phases; however, these levels of change can be recognized as the sequential phases of change implementation, implying that organizational change may experience different levels or phases. "Organizational change can be conceptualized as a continuum of relationships with three degrees or levels of changes, each with its clearly defined domain, intended outcomes, and definite strategies. These levels are not discrete types, rather they represent a progression" (Kumar & Thibodeaux, 1990, p. 39). In other words, the organization change starts from the lower level, then moves upwards, and eventually, advances to the institutional level, which involves all aspects of the organization. In this sense, the ideal is for change to reach the highest level or the institutionalization phase since it is only at this level that an organization can completely renew and transform itself.

3.4 Strategies of Organizational Change

Given the question of how to implement organizational change, Chin and Benne (1985) proposed three approaches for change execution: empirical-rational, normative-re-educative, and power-coercive. The empirical-rational strategy is based upon a rational human being who can recognize the gains and losses of change. As long as the human being is persuaded by the change outcome, he or she is willing to support the change. This method represents the classical way of implementing organizational change where the organization establishes goals, designs the most appropriate process and allocates resources on the basis of full rationality. From this point of view, change can simply be treated in an objective manner and is reduced to a technological matter. However, the assumption is fallacious in two aspects: firstly, human beings are not fully rational; the rationality of humans is limited, so the gains and losses of change cannot be fully measured by human beings. Secondly, organizational change cannot be treated simply as a technological matter since it always involves values, norms and attitudes of the organizational members.

The normative-re-educative strategy recognizes the importance of the values and attitudes of the members in the organization and therefore turns to the cognitive method so as to shift the organizational members' cognitive structure and values into line with the organization's objectives. This approach has been widely documented in the literature (Kotter, 1995; Moesby, 2004; Cohen, 2005) as a way of stimulating the staff members in the organization. In higher education, the normative-re-educative strategy is frequently used to bring about educational change because it can be used with a large number of staff development programs which are intended to impart the needed knowledge and skills to the staff members. The basic assumption of this strategy is that the organizational members will be willing to participate in the change as long as they recognize the benefits of the change. It is believed that the normative-re-educative strategy is suited for long-term effectiveness since it aims to set the conditions for change to evolve gradually, rather than gain immediate results (Kolmos & De Graaff, 2007). The power-coercive strategy refers to the change process that is mainly enforced from the management level, regardless of the willingness of the organizational members. Though it might suit the emergent situations and needs, in the long run, it cannot maintain the momentum at the ground level. It is especially so when change happens in higher education institutions replete with professional staff members. The power-coercive strategy is often associated with a top-down approach where change is primarily initiated by the senior managers. As Kolmos and De Graaff (2007) noted, though a power coercive strategy is effective in dealing with emergent problems, the top-down style change is unlikely to sustain change as it lacks recognition and support from members of the organization.

It can be argued that all strategies are flawed when it comes to the understanding of the human factors involved in the change process. The power-coercive strategy assumes that power is the dominant impetus to propel change and thus pays significant attention to the use of power itself rather than the human beings in the organization. The empirical-rational strategy, with the assumption of rational human beings, has the tendency to reduce the change process to a technological matter. In both cases, change can be seen as completely objective and aspects of the proposed change be readily measured by human rationality. In this sense, change becomes objective and neutral and thus diminishes the importance of human values and attitudes in the change process. The normative-re-educative approach notes the importance of the individual's values and attitudes in the change process and aims to alter them through knowledge transfer and education, i.e., staff development programs. However, this approach assumes a correct or valid value and attitude to the change process and sees it as necessary to transform the organizational members' cognition to the "correct" one. Due to its strong normative nature, the normative-re-education strategy fails to understand the organization members' subjective condition in the change process such as the staff members' values and worldviews. Thus, the normative-re-educative strategy is unable to answer or address questions such as how the organizational members understand the change objectives, strategy and process.

3.5 Motivating Members of the Organization

Recognizing the importance of the organization's members has been widely regarded as a crucial element in a successful transition to PBL organizational change. As noted by Kotter (1995), Moesby (2004) and Cohen (2005), maintaining momentum in the organization is always the key factor for a successful organizational change. The eight phases outlined by Cohen (2005) for implementing organizational change mainly focus on triggering and maintaining the motivation of the members of the organization by: creating a sense of urgency, forming a powerful guiding coalition, creating a vision, communicating the vision, empowering others to act on the vision, planning for and creating short-term wins, consolidating improvements and producing still more change, and institutionalizing new approaches.

There are several ways of addressing the motivation of the organization's members. Scientific management theory suggests the use of economic rewards on the basis of satisfactory or compliant behavior. The stimulus, therefore, primarily focuses on one's behavior and ignores the internal process of human beings; effectively this theory treats people as machines. The humanist theorists note that humans are motivated by far more than just economics and point to the human need of getting attention. The political and the collegial model of the organization capture the political dimension of human beings and emphasize humans' pursuit of their own power and interest. This theory suggest that the best way of motivating human beings in an organization is to pay attention to the power and interest needs of different organization members or different parties. Other theories emphasize the cognitive dimension of human beings and therefore propose that the "normative-re-educative" strategy (Chin & Benne, 1985) be used to align the cognition of the organizational members with the goals of the organization. The educative strategy of normative re-education is a way of imparting the necessary knowledge and skills to the members of the organization in the hope that they internalize the knowledge and feel motivated to support and participate in the change process.

3.6 Analysis of the Organizational Approaches

All of the preceding approaches to organizational analysis, change processes, levels and strategies are flawed as following:

Firstly, they represent a typical method of conducting structural analysis, largely relying upon the principle of dissection. The rationale of the structural analysis is to dissect the organization's structure and the transition process through a set of elements that can be explicitly identified, measured and controlled. Even the organizational anarchy approach, claiming to deal with the uncertainty and ambiguity of the organizational process, relies heavily upon structural analysis. This approach does not only directly adopt the notions of objectives, technology, participants, etc., as seen in Scott's work (1987), but also applies a quantitative method to analysis and predict the organizational process (e.g. Cohen, March & Olsen, 1972). This type of structural analysis also embodies the theories dealing with the organizational change phases, which dissect the change process into different stages, each with their own objectives, strategies and activities. The current change process theorists, capturing the time sequence traits of the change process, describe the organization change process as a series of planned phases, each with its particular mission, activity and expected outcomes. The strength of a structural analysis is that it facilitates a focus on a small unit of the organization at a time. Through the use of a variety of approaches, one is able to see many different aspects of the organization within the change process. For instance, the contingency theory assumes that an organization will alter its structure and performance according to changes in the organization's environment. Alternatively, the human-organization theory hypothesizes that the human factor is crucial to the fate of organizational change. As our knowledge of different aspects of organizational change increases, we may be able to formulate a more complete picture of knowledge by assembling various theoretical pieces of the picture.

One assumption of a structural analysis is that one can understand the entire picture by comprehending the aggregation of the parts of the whole. However, the whole cannot easily be reduced to assembling a collection of parts, nor can the aggregation of different change phases be equated to a coherent change process. These theories, at their best, depict organizational change as a set of successive activities or events. We should be aware that an organizational process is far more than assembling a set of sequential activities. Therefore, it is safe to draw the conclusion that all preceding approaches of the organizational analysis from the bureaucratic model, human-organization perspective, open-system approach, or phased model, have limitations when it comes to viewing the organizational change as a process. They represent a rather static organizational profile.

Furthermore, the preceding approaches to organizational analysis involve a normative dimension rather than seeking to "understand" the organizational phenomenon. They hypothesize on the correctness and appropriateness of certain kinds of organizational ends and processes with which the practitioners need to comply. Whether it is the bureaucratic model, human-organization perspective or open-system theory, they all prioritize the effectiveness of organizational action, albeit through the use of different means. The bureaucratic organization approach might resort to formal authority, explicit objectives, task distribution and the execution of organizational rules, while the human-organization perspective takes the traits and the needs of the individuals into consideration; yet they both presume that human factors should be directed, guided and manipulated to comply with the organizational objectives, ranging from political statements (such as consensus and shared power) to cognitive articulations (such as "building a shared vision"). Both perspectives attempt to control, reduce and eliminate the divergence of the human factor from the organizational objectives. Since human beings are not passive objects who conform to organizational requirements, the attempts to normalize the behaviors of the organizational members are always unsatisfying and create constant and significant tension in the organizations. For example, Jaffee (2008, xvii) pinpoints the common tensions in organization theories due to their intention to control and manage human beings in an organization.

A further problem associated with the normative approach is that although it can be employed to design the organizational objectives and processes, it is flawed in the understanding of the organizational members' subjective world. Since the conceptions of the actors have a significant impact on the organizational change (Henriksen et al., 2004), we cannot ignore the importance of the worldviews of the organizational members when dealing with the organizational process.

4

Methodology

When choosing research methods, researchers should consider how well the methods fit the purpose of the study. The organizational members must be able to understand and interpret PBL implementation in higher education institutes and how these interpretations influence the change process. To this end, we need to select a research approach that takes into account the detailed organizational process and the human's subjective understanding of the organization. Therefore, a case study is employed as the basic strategy for this study. In this section, we explain the choice of research strategy, the data collection method, our methods of data presentation and analysis and considerations related to the quality of the research.

4.1 Case Study Strategy

The case study has been widely discussed in various research domains (Yin, 1994, 2003; Flyvbjerg, 2001, 2006; Cohen, Manion & Morrison, 2007). The case study is usually described as "an investigation to retain the holistic and meaningful characteristics of real life events—such as individual life cycles, organizational and managerial processes, neighborhood change, international relations and the maturation of industries" (Yin, 1994: 3). Unlike the quantitative research approach, which uses a large quantity of samples, the focus of a case study is usually on one or a small number of units.

We have chosen the case study as the main research strategy for this study based on the following considerations:

(1) This study is concerned with an organizational process that happens in natural settings (Creswell, 2009) in higher education institutions, which makes it impossible for the researcher to control and manipulate the situation. The organizational process cannot be brought into a lab. Moreover, the real-life process cannot be separated from its context. Therefore, a case study that can explore a social phenomenon within its real-life context (Yin, 1994: 13) is the best fit for this study.

(2) A case study can reveal the complexity of a social phenomenon. Social processes, with their minutiae and contradictions, defy the simplification and reduction required in quantitative research. A case study, since it allows a thorough exploration and a rich account of an event or process, is more likely to capture the complexity of a social phenomenon.
(3) A case study allows for the use of multiple sources of data (Yin, 1994; Creswell, 2009). Unlike a quantitative research, which heavily relies upon a single source of data, case studies study favor the use of various data sources, such as documentation, interview and observation, in addition to quantitative data. The use of multiple data sources has many benefits. First, different data sources can complement each other. For instance, observations can document live events that cannot be captured by documents and interviews. Second, different data sources can cross-validate each other (a point we will return to later).
(4) Case studies are especially powerful for the production of context-dependent knowledge (Flyvbjerg, 2006), which is of significant value to various fields. PBL implementation, since it is embedded in the concrete organizational setting, cannot be separated from its context. Therefore, the case study as a means of yielding contextualized knowledge is suitable for the investigation of PBL implementation.
(5) Case studies promote the development of professional expertise in the researcher. As noted by Flyvbjerg (2001, 2006), a beginner largely relies upon the rule-governed use of analytical rationality, whereas a skilled expert's performance is based on tacit skills. The latter cannot be achieved by acquiring context-independent knowledge, such as a set of abstract rules, since it brings the learner to the beginner's level. Rather, tacit skills can only be developed through practice in real-life situations and reflection on the relationship between the knowledge thus gained and its context. As this is precisely what a case study entails, conducting one contributes to the researcher's development of tacit skills.

4.2 Data Collection

4.2.1 Selection of Cases

The selection of cases, or "sampling," has been widely discussed (Yin, 1994; Miles and Huberman, 1994; Marshall, 1996; Flyvbjerg, 2006; Seawright and Gerring, 2008). Qualitative sampling differs significantly from quantitative sampling, which aims to select the most representative samples from a population. The rationale of qualitative sampling is to reach the optimum fit between the selected case and the research purpose. The sampling is valid as long as it addresses the research questions. Sampling methods include theoretical sampling, purposive sampling and random sampling (Marshall, 1996). Although sampling is an important step for any research project, it is highly controversial. When a particular case is the focus of the research (an intrinsic case study), as opposed to instrumental casework, which requires the selection of a case from a number of alternatives, the problem of sampling does not exist (Seawright and Gerring, 2008). Nevertheless, the selection of the case must still be done carefully.

This study is concerned with two universities that for the past few years have been in the process of transforming their educational system from Lecture-Based Learning (LBL) to PBL. In both universities, the change from LBL to PBL has been far from smooth. The transformation process is replete with tensions and challenges. To understand them, it is crucial to explore how the organizations' members understand and interpret PBL implementation. Furthermore, both universities implemented PBL at the organizational level, excepting only a few courses, with the intention of making an organization-wide change. This brings us to the organizational dimension we mentioned in the problem statement section. For these reasons, these universities make suitable case studies for addressing our research questions.

Furthermore, as suggested by some researchers (Curtis, 2000), one of the criteria for case selection is that the case must be likely to generate rich information on the type of phenomenon studied. Due to the efforts of my research group and the cooperation of the two universities, I was given total access to the two organizations, which enabled me to communicate with their staff members as well as the students, to directly observe their lectures and group work and to collect and read their documentation. The easy access to the two universities enabled me to collect a large amount of data, which provides a solid base for me to make a rich and complete account of both cases.

The universities will be introduced in the case story section.

4.2.2 Official Documents

Official documents serve as a typical and repeatable source (Yin, 1994) for the basic facts and figures of an institution. Furthermore, they offer the official interpretation of events within a particular field. The official documents collected in this study cover:

Policy statements for curriculum change from the university and the department, programs and curriculum plans, staff documents relevant to

PBL implementation (such as staff application forms for PBL funding), student portfolios, university and curriculum brochures, evaluation reports for PBL implementation, research papers from both institutes and individual researchers, etc.

4.2.3 Interviews

The interview has been widely discussed in the literature (Wengraf, 2001; Yin, 2004; Cohen, Manion & Morrison, 2007; Creswell, 2009). The interview is especially powerful when researchers explore a subject's experience and his or her understanding of a particular event or phenomenon. There are various types of interviews, such as conversational interviews, the interview guidance approach, standardized open-ended interviews and closed-ended interviews (Cohen, Manion & Morrison, 2007); semi-structured interviews and group interviews (Bogdan & Biklen, 1992). In this study, we use the semi-structured interview form, which requires "a number of interviewer questions prepared in advance but such prepared questions are designed to be sufficiently open that the subsequent questions of the interviewer cannot be planned in advance but must be improvised in a careful and theorized way" (Wengraf, 2001: 5). As this quote implies, the interview should be viewed as a mutual construction between the researcher and the interviewee under the guidance of research concerns. Furthermore, it is open-ended in nature, which makes it possible to explore the interviewee's insights on a particular phenomenon or process in-depth, in contrast to the closed-end interview (Cohen, Manion & Morrison, 2007: 353), which permits the collection of information only according to the prescribed categories. In addition, semi-structured interviews allow sufficient flexibility and openness to approach different interviewees in different ways while still covering the same areas of interest (Noor, 2008).

In this study, I conducted a number of interviews at both universities, attempting to (1) understand how organizational members perceive and understand the transition to PBL; (2) collect factual information regarding PBL implementation which could not be obtained from other sources; (3) cross-validate the data alongside other data sources.
The targeted interviewees include:

- Senior managers: vice chancellor (president) for teaching affairs, faculty dean, departmental head, PBL director, etc;
- Administrative staff;
- Teaching and research staff;
- Technical support staff;

- Students;
- Others involved in PBL issues: external consultants, etc.

Numbers of interviewees of each type at the two universities are provided in the following table:

It is worth noting that the design of semi-structured interview guidelines is not an easy task. It requires several rounds of interview pretests, rephrasing and restructuring. I conducted most interviews by posing several predesigned open-ended questions closely linked to my study aims but not entirely strictly structured. The follow-up questions were improvised depending upon the context, the personality of the interviewee, and the interaction between the interviewer and the interviewee. For instance, the interviewer is responsible for initiating more questions when meeting a "silent" interviewee; on the other hand, when facing an interviewee who is attempting to control the conversation, the interviewer should be aware that their conversation needs to be "on track" and not deviate from the research objective.

My interview guide is as follows:

1 Could you please introduce yourself to me, explaining your educational background, your working experience, etc.?
2 What are the characteristics of PBL here compared with other educational methods? What do you think of PBL? What are the advantages and drawbacks of PBL?
3 How were you involved in PBL implementation? How do you implement it in your daily teaching and learning?
4 What are the reasons for your institute's change to PBL? What are your individual reasons for changing to PBL?
5 What are the challenges in implementing PBL?
6 What do you learn from practicing PBL? Is there any impact of PBL on you in terms of your thoughts and practice?

Type of Interviewees	XU	ZU
Senior managers	4	4
Administration staff		2
Teaching and research staff	6	19
Technical support staff	2	
Students	2	7
External consultant	1	
Total number	**15**	**32**

4.2.4 Observation

In observation researchers watch and record the behavior or the activities of an individual or a group of people at the research location. The benefits of observation have been discussed by many researchers (Yin, 1994; Cohen, Manion & Morrison, 2007; Creswell, 2009). One of the greatest advantages of observation is that it allows the researcher to "gather 'live' data from naturally occurring social situations" (Cohen, Manion & Morrison, 2007: 396). This "live" data on what is occurring cannot be obtained from other sources, and is usually the only available first-hand data, as documents and interviews provide only second-hand accounts.

For this study, observation offered a large body of useful data which I could not obtain from literature or interviews. For instance, the data regarding how PBL runs in the classroom and how students interact with each other is not fully reported in either the official documentation or the interview. The authors of official documents and interviewees tend to, consciously or unconsciously, omit or distort some information due to their subjectivity.

In this study, observation was employed to obtain data regarding:

- The space and resource arrangements for PBL, such as a group room;
- How PBL runs in large/small classes, different disciplines;
- How group work runs in different classes and disciplines;
- How staff members evaluate students within a PBL environment.

4.3 Data Presentation

The large amount of data collected in this study could be presented in various ways. I employ the narrative approach (Czarniawska, 1997, 1998), aiming at showing the detailed process of PBL implementation at both universities. Narration captures the causal relationships among different events. In the traditional scientific mode of inquiry, causality is built upon predefined theoretical propositions and hypotheses. In a narrative approach, causality is gradually constructed from the events on the basis of a narration, actors' understanding and interpretation, and the time sequence (Czarniawska, 1997, 1998). The narration is a mutual construction between the actors in the organization and the researcher. Therefore, the narration does not exclude the human factor from the data presenting process; rather, it encourages the involvement of human subjectivity in data demonstration, which thus facilitates the understanding of human actions.

4.4 Data Analysis

Data analysis always involves generalization[1] and reduction, such as organizing the data into different categories and patterns (Yin, 1994). This is widely recognized as a crucial step in case study, as it is in any other kind of research. However, summarizing is dangerous in case studies, as loss of detail can mean loss of value (Flyvbjerg, 2006), since the value of case studies lies in their rich accounts of real situations and context-dependent insight and interpretation. As to this study, we should admit that the data analysis only partially shows the richness of the case, and it generalizes only part of the case on the basis of the researcher's subjective cognition. The real-life situation offers far more than it is possible to show in the data. In order to reduce the limitations of generalizing the rich case story, we follow Flyvbjerg's (2006) suggestions: first, the complexity of the case, including its many different or sometimes even conflicting facets, should be demonstrated, and second, the case should not be linked to any specific theory; rather, the case should be related to a broader perspective across various theories.

4.4.1 Analytical Framework

This study primarily focuses on the organizational members' understanding of three major themes in the PBL implementation process: PBL in general, social learning and human motivation. These themes were chosen due to the following considerations:

First, PBL is the key concept of this study. Currently, since there is no general consensus on the definition of PBL, due to the existence of many variations and forms of PBL (Savin-Baden & Major, 2004), the concrete profiles of PBL rely largely upon the manner in which the educational theorists and practitioners interpret its uses, objectives, processes, etc. In this regard, it is crucial to investigate how the organizational members generally interpret PBL, since their conceptions of PBL will produce a significant impact on the transition to PBL. In general, PBL differs from LBL in three respects: problem centeredness, interdisciplinary learning and social learning (Kolmos, 2008). Problem centeredness means that the learning process and the curriculum design start with encountering the problem rather than with subject-based knowledge. The aim of problem centeredness is to link theoretical knowledge to real-life experience. Interdisciplinary learning is concerned with the

[1] Generalization in case studies is not the same as generalization in quantitative studies, as will be explained in the following section.

curriculum design process (coordinating different subjects), or the learning process, in which students are encouraged to work on the problem drawing on theories and the methods from various disciplines. Social learning refers to an approach where students work on the problem in groups, which is further elaborated upon in the following section.

Second, the interpretation of PBL mainly refers to how organizational members interpret PBL at the macro level, and, most importantly, it is primarily concerned with the educational design process. Therefore, interpretations of PBL provide quite limited insight into how students' learning activities actually take place at the micro level, such as the classroom or the group. The reason for making social learning an independent section in this study primarily lies in the fact that it is through the understanding of social learning that we can approach how PBL actually performs in reality at the micro level. In addition, social learning is recognized as a major trait of PBL that builds its theoretical foundation on the basis of social constructivism (Piaget, 1953; Vygotsky, 1962; Savin-Baden & Major, 2004). Through the preceding theoretical analysis we can see that social learning in this study does not only refer to the interaction among students, but also includes the tutor/facilitator, who can be seen as a part of the student's constructive process. Moreover, the inclusion of the tutor in social learning does not only refer to the tutor's assisting/facilitating role in students' learning, but also points to the restructuring of the relationship between the student (learner) and the authority (teacher). The restructuring of the student-authority relationship does not only shift the traditional oppressive relationship (as argued by Freire) to a relatively more equal, dialogical relationship, but also points to the emphasis of the student directive role in the learning process.

Finally, the organizational members' motivation is widely regarded as being crucial to the organizational change process (Cohen, 2005; Moesby, 2004; Kolmos & De Graaff, 2007). The momentum of the organizational members becomes extremely important when the organization is willing to introduce a long term change. In this study, we are concerned with how the organizational members at two universities are motivated in the PBL implementation process, and what factors in the organization demotivate the staff members and students from participating in PBL activities. It is in this section that we start to shift our attention to organizational aspects, which are intertwined with the change process.

It is also necessary to consider how to deal with these different interpretations of PBL implementation. A normative approach intends to execute managerial attempts and thus intrinsically to restrain or even eliminate interpretations that are not consistent with the managerial attempts. An approach that is weak in the normative dimension would encourage organizational members to develop their own interpretations and make their own innovations. Therefore, we need to reflect on our approaches to managing an organizational change as well as their possible influences on the organization.

4.4.2 How to Conduct Data Analysis

This study will use discussion of the conventional wisdom regarding PBL, social learning and organizational members' motivation as the benchmarks for data analysis. It is necessary to contrast and compare how the organizational members understand these three elements with what we find in the established theory. Roughly, this kind of analysis follows a normative method where an appropriate value or norm is established beforehand. However, we should express caution on the limitations of this method. On the one hand, it is necessary to use the normative approach because any kind of PBL implementation should be carried out in accordance with the basic principles of PBL. Without the essence of PBL, any organizational change is likely to fail. On the other hand, seemingly inappropriate or incorrect understanding and practice of PBL should not be automatically regarded as something to be eliminated, since these might present new possibilities for implementing PBL.

The data will be analyzed with the following considerations:

- How do the organizational members interpret PBL through the means of narration and practice? What are the commonalities and differences in how problem centeredness is interpreted in theory and in practice? What reasons may be responsible for this difference? What are its implications?
- How do the organizational members interpret social learning through the means of narration and practice? What are the commonalities and differences in how social learning is interpreted in theory and in practice? What reasons may be responsible for this difference? What are its implications?
- How do the organizational members interpret motivation through the means of narration and practice? What are the commonalities and differences in how motivation is interpreted in theory and in practice? What reasons may be responsible for this difference? What are its implications?

In addition to the analysis of the organizational members' interpretations of PBL, social learning and motivation, this study will reflect on our current approaches to managing these interpretations. In this respect, we are concerned with two approaches to organizational change, respectively, the normative approach, and the non-normative approach, which is weak in the normative methodology. In addition, the influence of these approaches on the organization will also be discussed in detail.

4.5 Research Quality Considerations

The issue of quality is critical for any kind of research, especially for case studies, as this type of research has received many critiques (Yin, 1994; Johnson, 1994; Flyvbjerg, 2006; Cohen, Manion & Morrison, 2007) that constitute significant challenges to the process and the product of case studies. The process of case studies can be obscure, as it largely relies upon the researcher's subjective operations, which makes it quite difficult for future researchers to follow. The findings of case studies are low in generalizability and are therefore unlikely to be applied to other settings. Given these critiques, it is necessary to elaborate upon how the quality of case studies is assured, both in general as well as in this study.

The quality of quantitative research within the paradigm of positivism is often connected to the notions of validity, reliability, generalizability and objectivity. Given the nature of qualitative research, which includes case studies, Healy and Perry (2000) assert that the quality of a study should be judged on the terms of its own paradigm rather than from a positivistic paradigm. Guba and Lincoln (1985) propose the use of the terms credibility, dependability, transferability and confirmability as the criteria for judging the quality of qualitative research. Furthermore, many different types of techniques have been developed to evaluate the quality of qualitative research (Jick, 1979; Guba & Linclon, 1985; Yin, 1994; Shenton, 2004; Flyvbjerg, 2006; Cohen, Manion & Morrison, 2007; Creswell, 2009). This section will address these terms and the relevant quality-assurance techniques one by one.

4.5.1 Validity

In its original sense, validity refers to whether an instrument can measure what it is supposed to measure (Golafshani, 2003; Cohen, Manion and Morrison, 2007). It should be noted that the validity of a case study is not the same as the validity of a quantitative study (Creswell, 2009). The validity of a case study refers to the practice of the researcher adopting certain procedures to check the accuracy of the findings (Gibbs, 2007). Guba and Lincoln (1985) use the term credibility to replace the notion of validity in qualitative research, and focus on the degree to which findings make sense and are reliable from the viewpoint of the readers. High validity normally implies that a study is highly convincing. Validity points to the consistency between the research and the reality. Furthermore, Henriksen et al. (2004) argue that validity can be assured by the constant interaction between the researcher and the actors. In a research study that involves the researcher's participation or interaction with the actors in the case, the validity "is secured by the contact between actors and observant, by the ability to jointly develop realities and to jointly conceptualize the changes that they have experienced" (p. 180). Validity in this sense also implies practical use, which means that the research can be considered valid as long as it assists the actors investigated in the research to develop their own thoughts about the reality, which may further change and shape the reality.

As for the case study, Yin (1994) and Creswell (2009: 191–192) propose a set of strategies to enhance the validity of case studies. We will not attempt to make a thorough exploration of all these strategies, but rather to specify those with relevance to this study.

The triangulation method (Jick, 1979; Yin, 1994, 2003; Cohen, Manion & Morrison, 2007; Creswell, 2009: 191–192) is to compare and contrast the data from different sources so as to build up a consistent and coherent justification for the research. It is a way of addressing one matter by studying it from different standpoints, which offers more convincing results than the use of a single method, since data can be biased or distorted by the researcher's view of the reality being investigated. There are many types of triangulation, such as data triangulation, theoretical triangulation, methodological triangulation and investigator triangulation (Patton, 1987). In this study, I am primarily concerned with data triangulation, in which a certain theoretical claim is proved and checked by different data sources. As for case study, the use of multiple data sources always contains a crosscheck dimension.

Member checking (Creswell, 2009) refers to taking the report or a specific part of the study to the participants to check the accuracy of the research. In this study, the interview by nature involves being checked by the participant because the interview is an interaction between the researcher and the participant. This serves as a partial means for the researcher's subjective bias to be corrected by the participants. The interaction between the researcher and the actors as a source for research validity has been discussed in the field (Henriksen et al., 2004).

Rich, thick description (Creswell, 2009) of a case can add to the validity of the research since it can deliver the detailed settings, processes and perspectives of a particular phenomenon to the reader. In this study, a genuine record of the PBL implementation process, the diverse perspectives about the PBL implementation from the organizational members and the tensions between them make the research more convincing.

4.5.2 Reliability

In quantitative research, reliability means the repeatability of the research, or in other words, whether another researcher can obtain the same result at a later date when he rigorously repeats the research procedures of the original researcher. In contrast, in qualitative research, the reliability of a case study refers to the reliability that the researcher's approach is consistent across different researchers and research settings (Gibbs, 2007). The consistency of the results, rather than the repetition of the same results, is the central focus when considering the issue of case study reliability. There are various types of methods to assure the reliability of a case study. As reliability entails a high level of publicity and repeatability of research procedures, Yin (2003) asserts that the procedures of the case study should be recorded in as much detail as possible. Gibbs (2007) outlines several steps for reliability procedures. The basic premise is to avoid inconsistency of the definition/meaning of the same term, code, concept in the same research. Therefore, the crosscheck is especially essential to ensure the reliability of the case study.

4.5.3 Generalization and Subjective Bias

In addition to reliability and validity, a further issue is the generalizability of the case study, since case studies are often challenged for their low level of generalizability (Johnson, 1994; Yin, 1994; Flyvbjerg, 2006; Cohen, Manion & Morrison, 2007). To some extent, the criticism about the case study's generalizability is not a large concern, since the essence of a case study is in its particularity rather than its generalizability. When considering generalizability, Yin (1994, 2003) suggests that it is analytic generalization rather than statistical generalization that distinguishes the case study from quantitative research. Statistical generalization means that a sample represents the characteristics of a particular set of the population, and therefore the findings from the sample can be applied to the whole population. The notion of statistical generalization is pervasive in quantitative research, and some researchers also use the logic of statistical generalization to secure the generalizability of the case study. Dilverman (2000) argues that generalization of the case study can be realized when it is used in conjunction with other methods, such as quantitative measures of population, purposive sampling, and theoretical sampling or analytical models assuming the presence of generalizability in the existence of any case. However, the analytic generalization from Yin (1994: 10) and Curtis et al. (2000) means that the findings from a case study can be generalized to the sets of theoretical propositions. Therefore, researchers of case studies should generalize the findings to theoretical arguments rather than to populations. By contrast, Flyvbjerg (2006) proposes that a case study can achieve the aim of making generalizations when it is associated with the selection of a critical case. Further, generalizations can also be made from a case study through the "reasoning of falsification" (Flyvbjerg, 2006), which means that if a proposition is disproved by a particular case, then it is no longer valid in general.

Another challenge for case study is the researcher's bias (Cohen, Manion & Morrison, 2007), since data collection methods such as interviews and observation and the data analysis in a case study are heavily dependent upon the researcher's personal experience and expertise. "Bias" can also refer to the tendency to verify researcher's preconceived notions (Flyvbjerg, 2006). There are ways to minimize or even eliminate the subjective bias of both researchers and interviewees (Cohen, Manion & Morrison, 2007: 150). However, subjective bias is prevalent in all research fields, if we consider the design of categories and variables of quantitative research, which is done on the basis of the researcher's experience and expertise. Compared with other research types, the case study leans less towards the verification of the researcher's preconceived notions than other research types do (Flyvbjerg, 2006).

From a hermeneutical viewpoint, this study sees subjective bias as the condition of our understanding (Gallagher, 1992). Knowledge is always based upon humans' subjective construction. There is no research which can be absolutely independent of the human's subjective participation. It is so even for the natural science research which claims the pure objectivity of the research. This study tends to treat bias in its positive manner. By nature, the research methods used in this study, such as the interview, rely upon the interaction between the researcher and the actors (Wengraf, 2001; Henriksen et. al., 2004), which implies that the research is, to some extent, a mutual construction between the researcher and the actors in the case. In this sense, the researcher's bias cannot be completely removed from the research process. Again, we come back to the point that bias does not only enable our research, but is an integral part of our research process that constructs new knowledge. Last but not least, if we refer to the reliability and validity issue mentioned above, we see that the researcher cannot conduct the research in an arbitrary manner. Rather, the researcher's subjective bias is largely confined by his consideration of the investigation's validity and reliability. The rich account of the research process, mutual interaction between the researcher and the observants and cross-validation by different data sources combine to show that, although case studies involve a large amount of the researcher's subjective participation, it does not mean that the researcher can investigate or draw conclusions in any way he chooses.

5

PBL Implementation at X University

X University (abbreviated XU), historically an engineering teaching institution, is currently one of the largest, most multi-sectored and most culturally diverse higher education institutions in Australia, with over 50,000 students (including students at international sites) of over 100 nationalities on 12 different campuses in and around Melbourne. A large number of the international students are self-financed and as such they have to pay high tuition fees to the university. The main campus of XU is located in an old industrial and less developed region in the suburbs of Melbourne. Since its inception, XU has had a strong mission of contributing to the local community and beyond through networking with local, regional, national and international partners.

In 2006, the Problem Based Learning (PBL) approach was introduced and implemented in the School of Electric and Electronic Engineering (EE) and the School of Architecture, Civil and Mechanic Engineering (ACME) in the Faculty of Health, Engineering and Science (FHEC). FHEC was the largest and most culturally diverse faculty at XU. Traditionally, the engineering programs at XU followed a conventional fashion of teaching and learning in terms of knowledge centeredness, single disciplinary learning and an individual learning approach. This conventional educational paradigm, which was labeled as Lectured Based Learning (LBL), has been used since the foundation of XU. However, due to managerial endeavors, since 2006, XU has witnessed a radical change in its educational approach from LBL to the PBL methodology, featuring problem centeredness, group learning and an interdisciplinary approach (Kolmos, 2008), marking a milestone in educational change at the university.

PBL was initiated by the senior manager, the vice chancellor at XU. Through powerful managerial propulsion and strong institutional support, the PBL curriculum was finally established and adopted across the two engineering programs, EE and ACME. Due to the institutional arrangement, EE and ACME followed different paths independently and eventually developed two distinct PBL models: a holistic approach integrating several subjects by a common large project coordinating different subjects (EE PBL), and a single disciplinary-based PBL approach without coordinating different subjects (ACME PBL). In 2008, due to organizational restructuring, the EE PBL and the ACME PBL models were replaced by a common PBL model. In spite of strong institutional support in terms of resources, personnel and policy, the transformation from LBL to PBL, and from the two different PBL models to the common PBL, was constantly met by tensions and struggles between different forces.

5.1 Background

The introduction of PBL into XU has been widely documented (Simcock, Shi, & Thorn, 2009; Stojcevski & Ozansoy, 2009; Li, Du & Stojcevski, 2009; Li & Henriksen, 2010). Although PBL was mainly initiated by the senior manager of the university in 2006, the PBL implementation should not be reduced to a simple managerial initiative; rather, it can also be linked to the national and organizational context and to individual teachers' concerns about educational problems at XU.

5.1.1 Graduate Ability

Today's economy and society have posed great challenges to the traditional engineering educational practice in higher education. The fragile world economy, the mobility of students and professionals, the use of communication and instructional technology and the increasing articulation of social needs have shifted the profile of engineering practices to require more than simply technical skills (Shuman, Besterfield, & Mcgourty, 2005), both nationally and internationally. It is therefore necessary to conduct a scrutiny of the engineering graduates' attributes so as to ensure they are in line with the needs of the society. In Australia today, graduates' abilities draw increasing attention from government and industry, as well as higher education institutions. A great many attempts have been made to address which skills are needed for the future workforce, such as Engineers Australia (1998), ACNielsen Research Services' (2000) nationwide survey concerning employers' perception of graduates' skills and the employability framework published by the Department of Education, Science and Training (2002). Engineers Australia, the professional body in the engineering domain of Australia, committed to the accreditation of undergraduate engineering programs, prescribed a set of graduate attributes

(DETYA, 1998: 47) that had a significant impact on the curriculum of engineering programs in Australia. The ACNielsen Research Services' survey identified a range of skills that employers seek when recruiting new graduates:

> "academic achievement in a suitable discipline, literacy, numeracy, basic computer skills, time management skills, written business communication skills, oral communication skills, interpersonal skills, team working skills, problem solving skills and comprehension of business processes"
>
> (ACNielsen Research Services, 2000:7).

This survey also identified a notable mismatch between employers' expectations of the graduates' skills and the graduates' actual performance after they enter the workforce. The skills that the graduates lack the most, from the perspective of employers, are: problem-solving skills, oral business communication skills, interpersonal skills and understanding of business practices (ACNielsen Research Services, 2000: 22, 41).

More influentially, the Australia Chamber of Commerce and Industry and the Business Council of Australia carried out a collaborative research project in 2001 to identify the skills necessary for success in industry and business. The results of this survey eventually became a well-known and frequently cited framework for employability skills:

- Communication skills that contribute to productive and harmonious relations between employees and customers;
- Teamwork skills that contribute to productive working relationships and outcomes;
- Problem-solving skills that contribute to productive outcomes;
- Initiative and enterprise skills that contribute to innovative outcomes;
- Planning and organizing skills that contribute to long-term and short-term strategic planning;
- Self-management skills that contribute to employee satisfaction and growth;
- Learning skills that contribute to ongoing improvement and expansion in employees;
- Company operations and outcomes;
- Technology skills that contribute to effective execution of tasks (DEST, 2002).

Instead of solely focusing on the technical skills of graduates, this framework includes many relatively "soft skills," such as communication, problem solving, team work, etc., shifting the engineering programs' focus from technical skills alone to including soft skills as well. This framework significantly impacted the policy and the behavior of a huge number of educational institutions, which started to incorporate these skills when designing educational objectives and course content. Ever since, specifying graduate attributes has been integrated into university policy as part of funding and reporting arrangements by government (Precision Consultancy, 2007: 12).

Within this national context, the graduates' attributes at XU became an institution-wide concern, since the university had traditionally focused on the delivery of technical knowledge rather than the cultivation of soft skills. A number of staff members noticed that XU students were "technically competent" but "lack the soft skills that a graduate engineer should possess" (Stojcevski and Veljanovski, 2009). To address this concern, in 2004 XU held a forum comprising 32 industry participants and 25 members of the XU staff to examine the implications of global economic change on the western Melbourne region. It once more emphasized the need to closely link its curriculum with industry and promote cultivation of soft skills, such as problem solving and interpersonal skills. XU called for a transformation of its teaching and learning system to assist students in gaining the desired skills for industry and employers.

5.1.2 National Grant and Funding

The governmental funding project served as another incentive for higher education institutes to carry out educational innovations in Australia. The Collaboration and Structural Reform fund (CASR, subsumed into the new Diversity and Structural Adjustment Fund in 2008) was created in 2005 to enhance teaching, learning and research outcomes by encouraging structural reform of higher educational institutions and promoting the relationships among universities, business, industry and the wider community (Commonwealth of Australia, 2006). In response to the request of the Association of Professional Engineers, Scientists and Managers Australia for governmental assistance in helping universities to retain and promote engineering courses, the Department of Education, Science and Training in Australia encouraged the universities to apply for CASR funds to restructure their engineering curriculum. This gave rise to the "Re-engineering Engineering" concept plan in 2005, for which XU successfully applied for and received a CASR grant of $299,000 spread over three years. As required by the terms of the grant, the university matched this funding level to support the establishment of PBL degree courses in engineering.

5.1.3 Attrition and Enrollment

The student attrition rate normally indicates the quality of teaching and learning of an educational institution. In Australia, the Department of Education, Science and Training used the student attrition rate as an indicator for funding allocation from the Learning and Teaching Performance Fund. Unfortunately, XU didn't perform well on this indicator, ranking 29th of 38 Australian universities in 2003 in terms of commencing domestic bachelor student inverse attrition rate (Gabb, Milne, Cao, 2006: 2). Concern about the student attrition issue at XU led to the production of several internal reports (e.g.: Cao, Gabb, 2006; Gabb, 2006; Gabb, Milne, Cao, 2006) aiming to identify the causes for student withdrawal and propose recommendations to resolve the problem. It appeared that multiple factors could be responsible for student attrition: previous academic achievements, course preference, course fit, teaching quality, student background (language, parental education, financial situation, commuting time) (Gabb, Milne & Cao, 2006: 6–13); progress rate, attendance type, region, broad field of study, basis for admission (Cao & Gabb, 2006:v). Many factors were far beyond institutional control. For instance, universities have a limited ability to intervene in students' backgrounds, prior academic experience or their course preferences. Therefore, one practical and possible solution for attrition reduction was to improve teaching quality by enhancing student engagement and involvement in the learning process. As XU noticed, this argument was also confirmed by previous research revealing that enhancing student engagement by incorporating active learning and collaborative learning could significantly reduce student attrition rates (Gabb, 2006).

Apart from the national and organizational context, the decline in enrollment numbers, alongside the attrition rate, became two major concerns for the engineering programs at XU. One senior staff member experienced the restructuring of the industry field and intensive unemployment in the 1990s, attributing high student attrition in engineering programs to the historical results of that frustrating period, saying, "there were not many engineering job prospects at that time, people started to leave engineering and to go to courses which have good employment aspects. That was an initial start, then probably that continued together. Even though now there are a lot of schools for engineering, but the trend continues."

Apart from the historical effect, the low interest from students in engineering learning could also be linked to the current societal setting. A senior manager in XU noticed the fact that students nowadays "prefer to learn law and business rather than engineering" since law and business students earn higher incomes than engineering students after they graduate. A senior staff member noted that students today were less likely to choose engineering as their course preference in Australia.

Furthermore, the decline in the enrollment of students in engineering programs (often accompanied by a high student attrition rate) threatened the survival of engineering courses and programs. As many staff members commented, if the trend of low enrollment and high attrition rate of engineering programs continued in the future, some engineering courses and programs were likely to be closed down due to the lack of students. Thus, the engineering programs needed to commit to reversing this trend. As one senior staff member stated, engineering programs needed to "attract more students, and keep them." Though the university could do little with regard to the social context, as Gabb (2006) commented, it was able to make engineering programs more interesting and attractive, which was likely to interrupt the trend of losing more students and reverse it. PBL seemed to be one practical option in dealing with these issues. Although, as the managers realized, making engineering programs more interesting did not necessarily indicate that students in high school would choose XU because of PBL, at least this intention could make students aware of the difference between the conventional educational approach and PBL, which perhaps offered them something special.

With regard to student attention in the class, a lecturer noticed that in engineering lectures, "the retention is low, the attention is low, the participation of students in the learning process is low." Students appeared to be less interested in learning mathematics and science compared to other subjects. Probably, some staff members thought that it was the boredom that destroyed students' interest in learning. Therefore, a great many staff members hoped that PBL could promote the student's engagement in learning engineering subjects.

5.1.4 Organizational Profile

XU positioned itself as A New School of Thought, making great efforts to build its own specialties. Previously, some researchers have argued that one of the motivations of higher education institutions for introducing PBL was the intention of enhancing their organizational profiles (Kolmos & De Graaff,

2007). Specifically, in 2005, Professor Peter Paarr proposed that one benefit of PBL for XU was "raising the university's profile in engineering education." The improved organizational profile would be conducive to resolving some problems bothering the university, such as a high attrition rate and the decline of student enrollment. As one senior staff member commented, implementing PBL could make XU look different from other universities and therefore potentially more attractive to new students.

5.1.5 PBL before its Official Introduction

Prior to the formal introduction of PBL into XU, PBL, or at least some elements of PBL, may have already existed at XU, as some staff commented. Therefore, although PBL seemed new to the university, it was not entirely alien to some of the staff members. For instance, a young staff member recalled his educational experience,

> "When I did my degree, because I did an art degree, we did problem based learning in art, which was that you are given a problem or a project, and then you work on it, so PBL has been taught in art for a very long time. Like drawing a chair."

Another senior staff member who had industry experience commented,

> "You cannot teach engineering without bringing in problems from outside. The same applies to law, to accountancy, to marketing".

These arguments viewed PBL as a prevailing phenomenon in all educational fields, and therefore some staff concluded that they were already practicing a degree of PBL without labeling it as PBL. As a senior staff member commented, "XU for a long time... has been involved in delivering part of the content through PBL."

In addition, there were sporadic individual explorations and innovations regarding instructional approaches at the university, some of them intrinsically linked to PBL in terms of problem identification, group discussion and task allocation. One staff member involved in teaching a tertiary education program has experimented with a two-week cycle of PBL delivered classes. He recalled,

> "I presented them with a typical higher education, tertiary education problem, and then we met face to face in groups, facilitating groups, so it was more like the medical PBL in Aalborg, project based, identifying learning issues, allocating tasks in the first meeting. And

> then we did not meet again for two weeks, and they located the resources and posted them on the website. And then we came back together. We spent the first half of the meeting, going over that together, finishing that problem, and then started the next one. That was much more like a medical type of PBL".

The School of Nursing and Midwifery at XU had worked for many years on case studies in a quite highly simulated environment through means of technology (e.g., a digital human being). Additionally, there were requirements to complete nine hundred hours in hospital- and community-based learning and simulated learning activities for nursing and midwifery students. For the Course of Osteopathy in the School of Biomedical and Health Sciences, students were required to devote a significant number of clinical hours to working through their patients' problems. These teaching and learning approaches of case/problem analysis, group work and real-life settings are the essential features of PBL in the medical field.

The managers noted these elements of PBL, yet they highlighted the fact that the implementation of official PBL at XU would be focused on engineering programs. As they stated,

> "You might say that (medical PBL) is problem-based learning as well. But when you come to talk to us about problem-based learning, we are going to focus on engineering because we say that it is a whole curriculum approach."

Due to these different viewpoints on PBL, many staff members considered the official PBL implementation as a managerial strategy. As a senior staff commented,

> "The university wanted to formalize this and make sure that we do it correctly."

This "formalizing" incentive from the managers wanting to ensure a correct PBL model indicated future tensions and conflicts that would occur throughout the university.

5.2 Change Process

5.2.1 Decision Making

In the middle of 2005, the vice chancellor of XU employed a consultant to examine the viability of PBL commencing in the first-year program in 2006. The findings state:

> "The introduction of PBL will require substantial effort and commitment, and offers major benefits in return. These include helping XU graduates attain a demonstrably higher level of capability, halving attrition rates, and raising the University's profile in engineering education."
>
> (Simcock, Shi and Thorn, 2009)

PBL seemed to be a better way of than lectures of addressing issues such as students' competence, attrition and university profile. Although several reasons could account for the introduction of PBL at XU, it was the leadership of the vice chancellor of the university that initiated the change process, a fact that was also widely recognized by the majority of the staff members.

In June of 2005, the vice chancellor of XU instigated the program of introducing PBL into its engineering program, and gave all staff members half a year to prepare everything for the change. Some staff members recalled the preparation period as being "rushed" and "hasty," having to fulfill all the commitments in a short time period, including designing a brand new curriculum plan. It was decided that the first-year engineering program had to be delivered in a PBL paradigm in the fall of 2006, and the following years would follow a phased rollout schedule, so that the entire four-year program would have completed the transformation by the fall of 2009. In July of 2005, the Board of Studies in the Faculty of Health, Engineering and Science approved the new first year curriculum proposal.

Throughout 2005 and 2006, a number of staff development programs, together with Central Queensland University, Aalborg University and other outside partners, were organized beforehand to assist XU staff in learning what PBL was and how PBL should be performed. In July 2005, a two-day seminar was held by professionals from Central Queensland University (CQU) presenting a wide range of PBL-related topics. From the 3rd to the 14th of October 2005, some professional experts from Aalborg University met with XU executives, academic directors and other interest groups, and recommended that the vision for PBL in XU should be called Project Oriented Problem Based Learning (POPBL). From the 24th to the 28th of July 2006, another workshop was held by professionals from Aalborg University, "pre-planning for a change towards POPBL," to assist XU staff in better understanding the principles and philosophies of PBL through lectures and hands-on project work. The external partners assisted XU in PBL implementation in two aspects: first, they assisted XU in developing a vision and a detailed curriculum plan, since XU had no previous experience in implementing PBL in an institution-wide manner; second, they imparted the necessary knowledge and skills related to PBL to the staff members at XU. As some staff members recalled,

> "Before the PBL introduction here I had no knowledge about what PBL is and how it can be done. Through these training activities I came to know a lot about PBL. That really helps!"

In September 2006, two engineering schools, the School of Electric and Electronic Engineering, and the School of Architecture, Civil and Mechanical Engineering, began to convert their first-year program structures to a PBL model, though in different ways, eventually formulating two different PBL approaches.

5.2.2 Curriculum

ACME's program consisted of four units (see figure 5.1): two purely technical subjects (physics and math), and two additional subjects (experimentation and computing and engineering profession). A total amount of half of the course content was delivered using a PBL approach (Mills & Treagust, 2003). For each unit, a certain number of PBL elements were appointed, yet the proportion of PBL varied among different units. Each unit had its own problem/project component, serving as a complementary entity for the subject it belonged to.

EE introduced a holistic approach (see figure 5.2), which led to a radical change involving the process of redesigning the whole curriculum system. Prior to the change, there were four segmented subjects in the curriculum system, as shown in Figure 5.2: math, physics, circuit theory and electronics and computer engineering. All of them were focused on knowledge acquisition. After the redesign process, the former two units were combined into one, designated as "Electrical Fundamentals," and the other two were merged into "Enabling Science". These two units retained lectured-based teaching, aiming to offer students fundamental knowledge of engineering and serving as a support to the project unit. The unit "PBL and Engineering Practice" was developed as the problem/project-based learning component, allowing students to work in groups and connect what they learned in lectures to real problems and practical situations. The ratio of subject units to PBL component was 1:1.

The assessment method of the PBL courses was a major issue in the change process. PBL calls for an alignment between its objectives and its assessment method. Therefore, in 2006, SEE utilized the Structure of Observed Learning

Outcome taxonomy to examine if students achieved the expected learning outcome in the taxonomy. In practice, it was a combination of formative assessment and summative assessment. Students were required to submit portfolios as summative reports, accompanied by some formative components as an appendix such as team reports and individual journals, which aimed to assess the learning process. However, this assessment method was quantitative-oriented and had some weaknesses in identifying individual contribution in a group. Therefore, in 2007, SEE altered its assessment method to facilitate supervisors to judge individual achievement.

5.2.3 Institutional Support

Undoubtedly, the change to PBL obtained enormous institutional support, especially from the management level. The senior manager was a champion of PBL, playing a vital role in this significant change. When recalling the PBL implementation process, most of the staff members stated that it was the vice chancellor that made the decision to implement PBL in engineering schools at XU. Not only did she try to persuade her subordinates and colleagues to engage in this transformation process, but she also "pushed" the organization to advance when the change process was at a standstill. The deans and the heads of schools were equally enthusiastic about PBL. They described PBL as a "fabulous" educational paradigm. The creation of the Office for PBL highlighted the importance of PBL in the university agenda. A group of champions also formed to provide an impetus for the change.

Since PBL favored teamwork and group work, XU also made a rearrangement of its infrastructure to better advance the implementation of PBL. XU

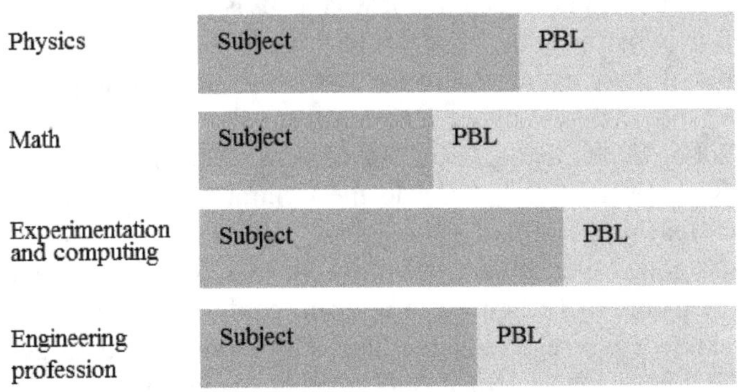

Figure 5.1 ACME PBL Model

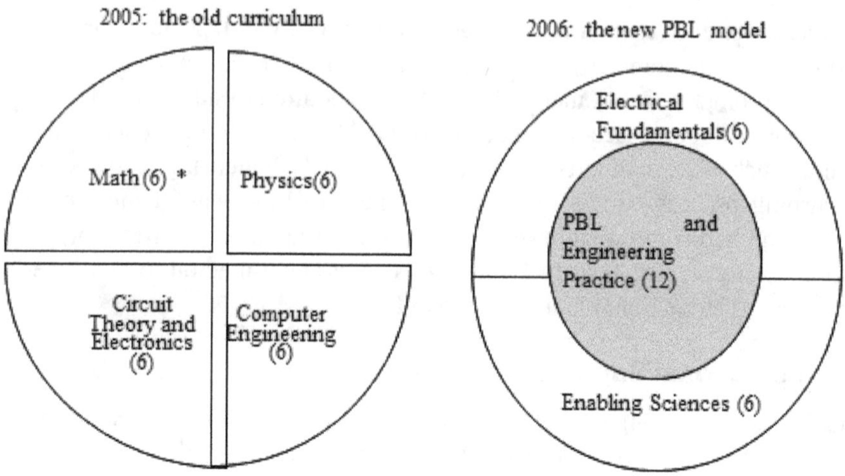

Figure 5.2 EE PBL for the First-Year Program(* the number indicates the credit points)

invested a considerable amount of financial resources into its space rearrangement. Large lecture rooms and experimental laboratories were remodeled to create PBL environments by adding whiteboards and PC access and dividing seat positions into groups. In addition, new PBL studios and group rooms were constructed to fulfill PBL needs, as shown in Figure 5.3. Finally, in order to make the learning environment more home-like, appliances such as fridges, microwaves, ovens, etc., were installed to enhance students' willingness to stay and work in PBL studios.

For teaching practice, the university arranged a considerable number of seminars, workshops, forums and training programs to offer staff more opportunities to gain a better understanding of what PBL is and how to implement PBL, as well as what benefits PBL would bring to teaching and learning. To name a few of these events, there were seminars or workshops by CQU professionals (July 2005), Aalborg University experts (September, October 2005, July 2006, September, 2006), XU experts (September, November, 2005). At the university level, XU had the Central University Innovative Learning and Teaching Hub, where the staff could learn new educational technologies, among other things. The university also encouraged and supported staff members to participate in relevant conferences and courses to update their knowledge regarding teaching and learning, and to get their work published. For instance, some staff members were supported to attend the MPBL courses offered by Aalborg University, and they were expected to become the future champions to promote the change process. It is worth

Figure 5.3 Group Room at XU

mentioning that the participation in staff development programs at XU was based on the staff members' own willingness and preference. There was no mandatory requirement for the staff to attend. The intensive staff development programs concerning PBL were arranged periodically throughout the whole change process, although there were more frequent events and programs in the early stages. As some staff members recalled, these staff development programs were quite valuable since they could equip them with PBL-related knowledge and skills.

The managers also referred to diverse resources available to engage staff in the change process, such as financial awards, grants and promotion, etc. There were different levels of grant opportunities for staff members (national, university and faculty). The university encouraged staff to apply for national funding options, one of which granted 50,000 Australian dollars (approximately a junior lecture's annual income) to the best teacher. There were also internal grants for teaching and learning at the university. Furthermore, XU and the faculty tried to mirror the national system to reward staff that contributed to teaching and learning.

Apart from the staff development programs and the financial support, managers emphasized the promotion policy to motivate staff since getting promoted was considered the "natural incentive for the staff". Since teaching was the main factor in appointment and promotion at XU, the staff needed to put a great deal of effort into their teaching quality to increase chances of promotion. This was recognized by a young member of staff:

> "There is no such policy or statement saying that if we get involved (in PBL), it will help your promotion. But of course, everyone knows it will, eventually, because the university is going through such a transformation."

5.2.4 Obstacles, Resistance and Challenges

5.2.4.1 Zoo Profile

In considering the challenges of undertaking such changes, a great many interviewees referred to staff resistance at XU. One senior manager sensed that this was "an unpopular change," stating that "many staff members don't know about PBL, and they don't want to know about PBL." The university offered intensive training programs in the early phase of the change process for the staff in order to help them to become familiar with PBL, as the majority of staff members at XU had no previous experience in PBL. However, there were a certain number of staff members who were quite reluctant to participate in these training activities. As one senior manager recalled,

> "They rarely showed up to the training program held by the faculty and the university."

An organizer of staff training programs noticed that some staff members did not participate in the training programs in teaching periods because they claimed to have to teach and that they had other high priority jobs, but they did not even show up when training programs were held in their non-teaching periods. It is clear that the staff development programs could only benefit those who were willing to learn about PBL. For the staff members who were opposed to PBL, these programs could only influence them in a quite limited manner, if not at all. It remained difficult to get a consensus from the staff on PBL and to get people into the same room together to think about their courses. A senior staff member recalled the resistance in a staff workshop,

> "While he (the trainer) was stating 'pedagogically, PBL could bring us these benefits...' some engineering staff members interrupted and asked 'have you ever taught engineering courses?' Then he replied 'no.' 'Then shut up!' The engineering staff said."

It seemed that some staff members with an engineering teaching background were quite confident about their own knowledge and therefore they were "*mentally stubborn*" and quite reluctant to open their minds to new educational paradigms.

There was a strong "individual culture" in the faculty with little teamwork and cooperation. A course coordinator noticed the lack of cooperation among staff members:

> "We require students to work as a group effectively. At present, we, the academics, do not work as a group effectively."

An internal survey asked the question, "what has been the most stimulating or supportive aspects of the PBL initiative so far?" Staff answers were recorded as "None," "Has there been any?", "Nil" and "Must have missed it." To the question "What major changes (if any) do you think you will need to make in the way you work?", some staff replied "avoid PBL—focus on research more than ever", "I am not afraid of those staff who say they don't like PBL and keep away, I am afraid of those who say 'they are in', but in reality they are not" and one answered, "resign, seek job elsewhere".

In a seminar, the manager in charge of the PBL implementation depicted the XU staff in the change process as a collection of different animals in the zoo:

> "First of all, there is the ostrich who as we all know, puts its head in the sand, working on the principle that if they can't see you then you can't see them. This doesn't seem to be a very good survival strategy, but fortunately, the ostrich can also run very fast.
>
> Then there are the moles who burrow underground and are hard to find. Then when they think the change is complete, they pop back up again.

> Tigers are powerful, or at least that's what they want you to think. Their message is this: mess with me at your peril. Go make your change elsewhere, little person.
>
> Dogs know that they are far more powerful in a pack. They seek one another out and attack en masse.
>
> Owls are wise and knowledgeable people. They sit up on their branches in their trees, pontificating and pointing down at the trivial world below. They know better than you and are not slow to point this out, as well as pointing out all the little faults in your change project, which is, of course, somewhat below them.
>
> And then there are the snails. Well, you know those old snails, they just go so slow. They creep along at, well, a snail's pace, and hope that you will leave them to their own devices.
>
> The most prolific animal in the resistance zoo in engineering at XU is the saltwater crocodile—also known as the only remaining living dinosaur—enough said!"

These "animals," though they seem different in their outlooks, shared some commonalities. That is, they were not willing to participate in PBL activities. The ostriches could work effectively but tried to avoid the change. The moles simply avoided the change by being silent. The tigers and the dogs were the most resistant ones, and they opposed the change in different manners: respectively, individually and collectively. The owls had their own knowledge of PBL and were fond of criticizing the others. The snails had quite low work efficiency and they wanted to be left alone. The crocodiles stuck to the traditional methods and did not want to make any innovation at all. The resistance mainly came from the academic staff, especially from older staff members, rather than the technical staff and students, who seemed to get along easier with the change process. According to some managers, it was the older staff members who were the most reluctant to change. Due to this huge amount of resistance, PBL implementation at XU was quite challenging.

5.2.4.2 Policy

Other staff members, though subscribing to PBL, complained about the policy at XU when PBL was being introduced. The preparation time for the change was, as some complained, "too little time, painful, in terms of redeveloping curriculum and course material." A great many staff members mentioned that PBL implementation created quite a heavy "administrative burden" for them in the initial stages of the PBL implementation, since they had to redesign the curriculum, rewrite subject and project guides and prepare PBL documents in a quite limited amount of time.

Some staff members commented that there was "little planning" and a "lack of deep analysis" from the senior management with regard to the change process, and "not enough time, nor resources, and not the right people." Allotted resources and time were not sufficient, either. There was a "financial wrestle" all the time. There were complaints about the distribution of financial resources. One technician thought that it made less sense to spend money on infrastructure rather than staff. One interviewee held that motivating staff was a difficult task at XU, as the pay scale in the educational system was not so differentiated, compared to that of industry or business. The staff at the same level got more or less the same income, and thus they had little economic motivation to put efforts into PBL.

5.2.4.3 Pedagogical Debate

There were strong ongoing pedagogical contentions as to the value of PBL among staff members. While the ideal of PBL could be seen from documents and instruction practices, many staff members worried about the knowledge coverage. This makes little sense from the perspective of PBL proponents; as one technician insisted, "PBL is not about learning detail." The justification for PBL highlights the cultivation of generic skills and learning motivation, rather than the knowledge scope. Other staff members also worried about technical skills, and argued that students might not be developing enough technical skills in the new learning environment. Some committed staff stuck to the conventional way of instructing students even after the change had officially occurred. They organized classes and lectures in addition to the PBL courses for students to gain knowledge coverage, even without getting paid. Although these debates and disagreements at the university might superficially be harmful for the change, some staff viewed them from a constructive perspective, stating they "actually lead us to having more understanding of PBL, and the staff who were not in agreement, and also where to change."

5.2.5 Feedback

EE PBL seemed to achieve more in the change process than ACME PBL, in terms of curriculum and staff morale, from the perspective of both internal and external observers. The School of Electric and Electronic Engineering viewed PBL as a holistic curriculum approach and therefore introduced a radical change by systematically restructuring and redesigning its curriculum and courses. In its new PBL model, the problem/project stood in the center of the learning process, whereas the subjects served a scaffolding role. In contrast, the School of Architecture, Civil and Mechanical Engineering kept its original form of several isolated subjects with a certain amount of PBL elements in each of them, rather than introducing a systematic approach. As to staff morale, many internal and external observers commented that there was more enthusiasm in the School of Electric and Electronic Engineering than in the School of Architecture, Civil and Mechanical Engineering, where there was more unwillingness, reluctance and resistance to change. In addition, there seemed to be too many activities or short problems/projects in the School of Architecture, Civil and Mechanical Engineering because all four subjects had their own problem/project activities, which probably created a heavy academic burden on the students and thus damaged their learning motivation.

5.2.5.1 Feedback from Other Faculties and Staff Members

There was evidence from interviews that other faculties (e.g., the Faculty of Business and Law) were impressed by and interested in the way the engineering faculty arranged its space when implementing PBL, and that perhaps they might be influenced by PBL in the engineering program at XU in other ways, rather than in a direct way.

Staff attitudes towards PBL seemed to be segmented, as shown in the zoo profile in the previous section. The champions regarded PBL as a powerful model for engaging students in learning and addressing the levels of graduate competence (e.g., teamwork, communication, problem solving) expected by the labor market and the professional body, Engineers Australia. Some staff noted "higher attendance, higher attention, higher participation" of students in PBL classes than in normal lectures. PBL also benefited the staff itself in terms of teaching competences and knowledge. The risk of being questioned by students in classrooms forced staff to update their technical knowledge constantly. However, doubters of PBL remained at the stage of questioning the value of PBL. Meanwhile, the result of PBL was not always satisfying, as one senior lecturer criticized the group performance of students, "they do not cooperate well even after four years' learning."

5.2.5.2 Students' Feedback

The feedback of students seemed to be optimistic. One student interviewee thought PBL helped him learn how to be a professional and how to mix with different kinds of people, which was impossible in the traditional lecture format. A school-wide survey of first-year students' response to their learning experience in industry and community noted "nothing but positive feedback from students" (Simcock, Shi & Thorn, 2009). The least positive piece of feedback was that students found themselves "committing too much time to this subject" (Simcock, Shi & Thorn, 2009). It seems that students welcomed this new style of learning, as one student wrote in the portfolio,

> "The problem given to us this semester may have been difficult in certain aspects, but the fact that it gave us a slight look at what was going to be expected of us further in the course and later in the workforce was helpful and taught us the different fundamentals of electronics, programming and how mechanics can be combined with electronics. The problem pushed our skills and abilities that little (sic) further. As in this particular problem, we were not only required to create a solution and build a model of an Automated Stacker, but we were also made to interact with a client and ensure his needs of the stacker were satisfied."
>
> (Simcock, Shi, & Thorn, 2009)

However, the actual practice of PBL was unsatisfying. For instance, since there were no strict guidelines and procedures for PBL from the senior managers, the quality of PBL depended largely on the individual lecture or supervisor. Meanwhile, the tension in the university between the old staff, the young staff and the management also created obstacles for practicing PBL. Therefore, there was plenty of room for improvement in the delivery of PBL. As a student maintained, "Problem-Based Learning could be done better".

5.2.5.3 External Stakeholders

The external industry and community that had been working together with students also responded positively. As one representative from Braybrook Community Centre said,

> "It is amazing; they are still working on the project even after they are supposed to have finished. I have never seen students so interested and involved; it is a lot different from my time!"

Another managing director of Meridian Cables said,

> "I have gained a great deal from this experience and I am really looking forward to working with you and more students again next year."
>
> (Simcock, Shi, & Thorn, 2009)

5.2.5.4 Attrition and Enrollment

Concerning the attrition rate and enrollment numbers, one interviewee commented that PBL had already produced positive effects on their retention rate, although enrollment had not improved. However, on the whole, there seemed to be a few relevant statistical numbers verifying that PBL had already had a drastic impact on attrition and enrollment in a positive way. After all, attrition and enrollment were influenced by multiple factors, not just teaching quality. It may take a long time to examine the impact of PBL on attrition and enrollment.

5.2.5.5 Organizational Profile

Even though there were quite a few documents empirically proving that PBL implementation has improved XU's profile, we can obtain some clues from XU's advertising material (e.g., websites), validating the fact that XU attempted to set itself apart from other higher education institutions by offering a rich description of PBL on its website, in its curriculum introduction and in its recruitment booklet. Some staff members recommended that XU do some marketing in secondary schools by telling them what PBL is, why PBL differs from conventional teaching, what group work is and showing them the products that students had made in PBL environments.

5.3 Change Continues

5.3.1 Background

XU initiated a development plan in October of 2006 under the name of Making XU Program, designing a set of objectives that XU would reach by the year 2016. The aim of this plan was to further "align our services over the next few years more closely with the changing needs of students, enterprises and communities" by restructuring over 700 of its courses and curricula. This program required at least 25% of the learning to take place in the workplace and community. This commitment at the university level also impacted the curriculum in the engineering program.

There were several influential events at XU in 2008 that interfered with PBL implementation. The School of Electric and Electronic Engineering and the School of Architecture and Civil Mechanical Engineering were merged into the School of Engineering and Science due to financial and interdisciplinary cooperation considerations and the declining number of enrolled students. After the merger, one major problem was the coexistence of two distinct PBL approaches in a single school. After scrutinizing the PBL implementation in the new school, the review panel concluded that it was unwise to keep two different approaches in a single faculty. Moreover, it was beneficial to bring the staff together under the umbrella of a unified model because the coexistence of different understandings of PBL in a single school would cause conflicts. As a senior manager said, "you cannot have two groups of staff, one understanding the model over here, the other understanding the model over there." Therefore, the new engineering school needed to redesign a unified curriculum to replace the two different approaches.

Secondly, XU initiated a sustainable plan to enhance the university's competitiveness in the coming years. Part of the plan involved the reduction of campuses, schools and disciplines, as well as staff numbers. The implementation of this new plan had a significant impact on the university and PBL, which will be discussed later.

The change of the existing curriculum can also be attributed to the university's concern with the students' backgrounds. XU had a large proportion of students from non-English speaking backgrounds and from families with a lower social and economic status. In 2005 about half of the first-year students in engineering programs at XU were classified as low social economic status, over half of them were non-native English speakers and 43% of them were the first in their family to attend college (Keating & Gabb, 2006). Furthermore, the students at XU had quite limited experience with group work before entering college. The students' deficiency in group work experience, associated with their relatively disadvantaged academic background, called for a more supportive learning environment in the first year of the engineering programs. It seemed that a 50% PBL-based programPBL in the first year was too demanding for them.

5.3.2 Curriculum

In 2009, it was decided by the vice chancellor that a common model of PBL would be established at XU in 2010 to replace the currently separate PBL models, which would be termed the second generation of PBL or 2010 PBL. Earlier in 2007, XU made five commitments to achieve a distinctive

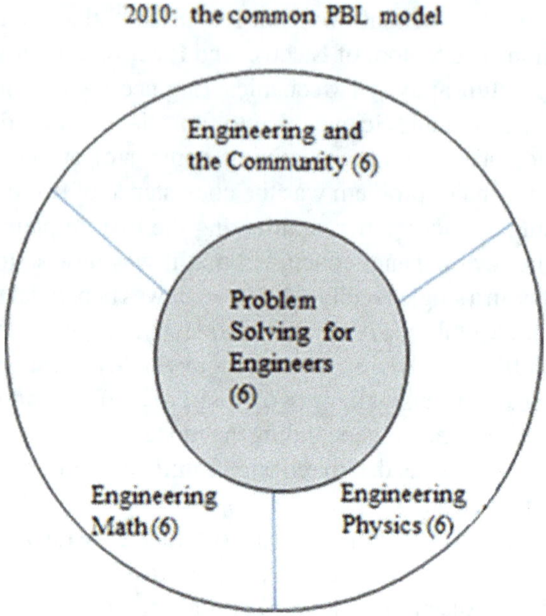

Figure 5.4 The common PBL model in 2010

and sustainable future at the university. One of the significant ones was the requirement for at least "25% Learning in the Workplace and Community (LiWC) a universal feature of XU courses" (XU, 2007). In line with this university policy, it was prescribed that 25% of the learning assessment would be based on workplace and community-based learning. Meanwhile, XU obtained another 1.5 million Australian dollars to restructure at least 25% of its assessable learning content into work-based learning (Collaboration and Structural Reform Fund, 2006).

The previous ACME PBL approach was totally abolished. The new PBL model was developed based on the EE PBL model, which also underwent drastic revision. Meanwhile, the new PBL model incorporated the Learning in Workplace and Community policy prescribing that 25% learning be in industry and/or the community. Thus the P in PBL extended its meaning to include "Problem" on campus, "Project" in the community, and "Practice" in industry. XU defined its new approach as P3BL, indicating a more contextualized and mature model for XU rather than one transplanted from other universities.

The new model (see Figure 5.4) continued the principles of the 2006 curriculum, with PBL at the center of the whole curriculum structure, with the lecture unit aimed at supporting the project unit. Since it was challenging for the students who previously had no experience of PBL to deal with 50% PBL in the first-year program, the new plan for the first-year program reduced the number of credits contributing to PBL to 25% of the total credits. "Engineering Fundamentals" was split into engineering math and engineering physics, returning to its original form (though not to its original content). The "Enabling Science" unit was moved to the second-year program and designated as only for the students whose specialization was Electric and Electronics. Two new units were developed: the new subject Engineering and the Community aimed to expose students to engineering ethics and responsibility issues; the PBL component Problem Solving for Engineers provided students with basic learning skills within the PBL context, such as how to describe a problem, how to conduct teamwork and how to design the solution to the problem. The students in the first-year program shared the same courses, but were split into different streams in the subsequent years. From the second year, students were able to design and solve problems within a community and/or industry context.

5.3.3 Challenges

The new change followed a top-down approach, mainly being enforced by the senior management and obtaining wide institutional support. Unavoidably, there were strong resistance and challenges. Prior to the establishment of the 2010 PBL model, the staff from EE and ACME had debates over the EE PBL and ACME PBL approach. EE people preferred one big, robust problem, covering as many elements and components as possible, whereas the ACME people leaned towards several smaller problems due to their discipline feature. Given this problem, the decision making for the implementation of the 2010 PBL was quite challenging, as a senior staff member recalled:

> "(The PBL director) set up a committee, and I joined soon afterwards. He... invited people to become members of this committee from anywhere. The course coordinator did not come, by and large, just some others came. That committee developed a model and presented it to the dean. The dean approved it, and then he left. Then the new dean [came]... The group of course coordinators came to her and said 'we completely disagree with what was produced, and if we are forced to do it, we resign.'... So we had a series of weekly meetings at least with course coordinators... to work a way through

this.. . . It took months and months and months of negotiation. . . but it was still imposed really against the will of the coordinator."

Since EE staff and ACME staff had different concepts of PBL, as mentioned above, the enforcement of the new common PBL model unavoidably caused significant controversy in the organization, both open and hidden. It took a long time to discuss and negotiate with opponents to get them to accept the plan of a common PBL model. Besides the pedagogical debate, there were other complaints about the policy. A senior lecturer thought it was better to consolidate the previous PBL model rather than simply abandon it and impose a new one.

In 2008, XU underwent a great many changes, such as school restructuring, uncertainty about the budget situation, staff reduction, the enforcement of the 2010 PBL model, learning in workplace and community policy, etc. The staff had to struggle with these competing agendas, as one interviewee perceived it, there were "too many background noises," "too many ideas floating around." In some cases, these changes were contradictory to the essence of PBL. Since the management had initiated the sustainable plan in 2008 with the aim of making XU more competitive, XU had found itself forced to reduce staff, courses and programs. In 2009, XU reduced its technical support staff in the School of Engineering and Science from 21 to 8. As one technician commented, PBL was "staff intensive," implying that PBL required an increased number of staff. The drastic reduction in the number of staff would definitely affect PBL implementation. Moreover, the enforcement of a sustainable agenda together with the staff reduction process lowered the morale within the institution because the staff's fear of losing their jobs pervaded the entire organization. A senior lecturer noticed that the motivation of staff was "a theoretical concept." This could in turn affect teaching quality at the university. The excessive number of changes was also likely to shift the university's focus to objectives differing from its original intention. For instance, the Learning in Workplace and Community policy played a vital role in the new PBL model, which made PBL itself look like a complementary component.

The personnel flow at the management level and the staff level added to the complexity and the uncertainty of the change process. The former dean and the director of the Office for PBL, who were both PBL champions, left the university and sought other positions in 2008 and 2009, respectively. The staff reduction process also caused technical staff loss: over half of the technical support staff in the engineering school was cut. The drastic personnel change, together with the excessive change agenda at XU and its low morale, created concerns about the future of the university and PBL. Some staff members were left "not knowing where the university would go," and wondering, "How is PBL going to function with less staff?"

Currently, the organizational change over to PBL at XU seems to have progressed into what is termed the "institutionalization phase" (Fullan, 2001), which implies that PBL has become an integral part of the university system. Although on a superficial level PBL has become a university routine with which the staff and the students have to comply, the undercurrents below the institutionalized PBL practice deserve the attention of both managers and staff members.

6

Organizational Change to PBL at Z University

6.1 Background

Z University, with 13,634 students (8,263 undergraduate students, 3,349 master students, and 1,028 PhD students), is located in Shenyang, the capital of northeast China's Liaoning province. Traditionally, ZU was a medical university with the mission of cultivating and producing future professional nurses, doctors and medical experts regionally and nationally. ZU has become a multi-disciplinary university covering humanities, social science, natural science, and engineering science, although medical education still dominates its teaching and research activities, and thus significantly shapes the university's public profile.

In 2004, ZU launched a large-scale teaching innovation project: transforming a large number of its subjects from LBL towards PBL across various disciplines throughout the entire university under the leadership of its vice president for teaching and learning. As PBL was implemented in many faculties and departments in a diverse manner at ZU, rather than in one single faculty or school as at XU, the introduction of PBL became dependent on both the unique character of each department and the character of the university as a whole. Therefore, a brief overview of the university's organizational structure and basic terminology closely relevant to PBL will precede further analysis of PBL implementation at ZU.

6.1.1 The Organizational Structure of Z University

The university has several units performing different functions (see Figure 6.1):

- The Office of Teaching Affairs is the university-level administrative body dealing with teaching and learning issues such as organizing the national curriculum or the university-level curriculum, planning classroom arrangements and other resource distribution, organizing curriculum evaluations and learning assessment activities, and superintending teaching activities, etc.

- The academic faculties and departments carry out teaching and research activities. At ZU there are over ten academic colleges and departments such as the College of Basic Medical Science, College of Pharmaceuticals, College of Public Health, College of Forensic Science, College of Nursing, Department of Information Management and Information System, Department of Bio-medical Engineering, Faculty of Social Science, etc. The largest among them is the College of Basic Medical Science, which performs the teaching and researching activities for fundamental medical sciences such as bio-chemistry and molecular biology, cell biology, pathology, histology and embryology, pathophysiology, immunology, etc.
- The Research Center for Medical Education carries out research activities relating to medical education and is one of China's best research institutes in the field of medical education. Apart from its research functions, this center also administrates teaching innovation and reform issues by planning, implementing, superintending and evaluating teaching innovation activities. At ZU, the vice president for teaching affairs also serves as the director of the Research Center for Medical Education, which is in charge of PBL initiation and implementation throughout the whole university. In addition, the Center offers one basic course, Introduction to Medical

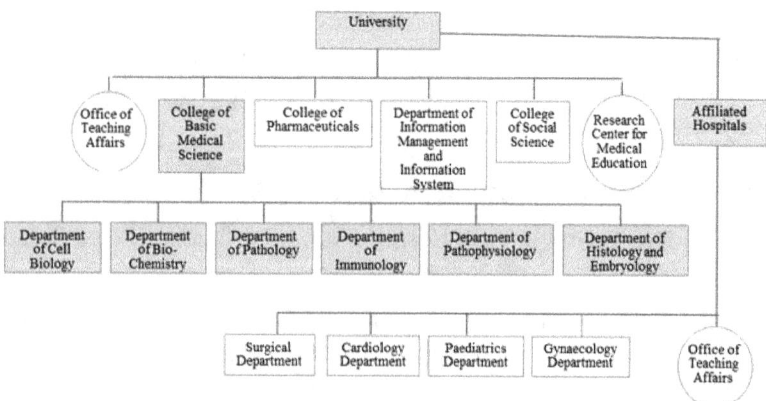

Figure 6.1 The Organization Structure of Z University

Science, to all undergraduate students in cooperation with the Office of Teaching Affairs.

- As a medical university, ZU has over ten affiliated hospitals throughout the country serving double roles including both clinical activities, like a normal hospital, and teaching and researching activities, like an institution of higher education. The structure of these hospitals partially mirrors that of the university, including an administrative unit, similar to the Office for Teaching Affairs, which is in charge of curriculum and teaching issues at the hospitals, and also including academic departments such as the Surgical Department, Cardiology Department, Pediatrics Department, Gynecology Department, etc., which conduct teaching and researching activities. However, the academic body of these hospitals significantly differs from the one at the university since the hospitals' clinical mission takes precedence over its academic mission.

6.1.2 Types of Curriculum

The formal curriculum at ZU is categorized into general education curriculum, fundamental medical courses and clinical courses.

- The general education curriculum, also called the national curriculum, includes the study of foreign language, natural science, humanities and social science, physical education and computer science. The Ministry of Education in China designed this curriculum as the minimum requirement for the graduation of university students. Roughly, it makes up one third of the total curriculum.
- The fundamental medical coursess are primarily offered by the College of Basic Medical Science at ZU for the students in lower grades, i.e. roughly the first six of ten total undergraduate semesters. These courses are intended to equip the students with basic medical knowledge and skills. Examples include Introduction to Clinical Medicine, Physiology, Bio Chemistry, Anatomy, Molecular Biology, Cytobiology, Pathology, Histology and Embryology, Immunology, among others.
- Clinical courses: The First Affiliated Hospital of ZU is primarily responsible for providing clinical courses to students from approximately the seventh semester to the tenth semester and arranging medical internship activities. Courses include Nursing, Pediatrics, Gynecology, Cardiology, Surgery, Topographic Anatomy, among others.

6.1.3 Types of Programs and Classes

It is necessary to define some terminologies used by the College of Basic Medical Science (CBMS). The CMBS serves Clinical Medicine major students, who form a majority within ZU. The CBMS encompasses 26 departments, which range from three academic staff members to over 40 academic staff members. The major responsibility of these departments is to design the relevant curriculum and conduct teaching activities in conformance with the regulations and guidance of the university. CBMS offers two different programs: the five-year (ten-semester) bachelor program and the seven-year (fourteen-semester) master program. The bachelor program took in an annual recruitment of around 1,000 students, in contrast to 300 students annually in the master program. To facilitate management, ZU organizes the students into classes, which can be further sorted into three categories:

- Normal small classes: containing 30 to 40 students. The five-year program is composed of 30 classes whereas the seven-year program consists of 10 classes.
- Normal big classes: four small classes comprise one big class. The size of a big class is around 120. At times, the teaching activities took place in the form of big classes when staff numbers were insufficient.
- Integrated classes: since 1998, ZU has embarked on an experiment called the Integrated Course Innovation Project, aimed at merging the large number of traditionally single disciplinary courses into a limited number of interdisciplinary courses. Eventually, both the five-year program and the seven-year program, respectively, designated four small classes as the official integrated classes in which students took interdisciplinary courses. This is in contrast to the normal classes where the students continued to take single disciplinary courses in the traditional fashion. For instance, the students from the integrated class took a course called Histology and Morphology of the Human Body as an integration of three different disciplines, Anatomy, Histology and Embryology, and Iconography. The students from the normal class were required to take Anatomy, Histology and Embryology, and Iconography as separate courses.

PBL was scattered across disciplines, departments, colleges, curricula, and classes. In the following sections, we will refer to these terms when recounting the story of PBL implementation at ZU and conducting our data analysis.

6.1.4 Curriculum Innovation Prior to PBL Implementation

Since the 1990s, ZU has witnessed several innovations aimed at enhancing its teaching and learning quality through changing its education and curriculum structure. From 1991 to 1996, ZU invested in the Assessment of Cognitive Skills Project with the financial support from CMB (China Medical Board), an independent American foundation for advancing health in China. The outcome of this project was a Question Bank, which later became a national medical examination question bank that is now used by over 95% of all medical colleges in China.

In 1998, ZU initiated another plan, the New Curriculum that attempted to integrate various isolated subjects. The rationale for integrating courses was drawn somewhat from the US experience. The vice president for teaching and learning at ZU travelled to the US for a long time period, and was quite impressed with integrated American medical courses. After he returned to China, he introduced this new concept into ZU. The realization of this conception resulted in changes in the course structure, textbooks, learning hours and assessment. For example, cytobiology and biochemistry were merged into one single subject: cell biology and chemistry. From 1998 to 2001, ZU integrated seventeen subjects into nine. This innovation led to the reduction of teaching hours. Notably, integrating courses did not necessarily involve integrating relevant departments, but rather coordinating those departments to deliver common courses together while still keeping their own departmental independence. ZU also designed and published several textbooks to better serve the integrated courses. Some courses also reduced their curriculum hours. For instance, the total teaching hours required for biochemistry dropped from 76 to 66. The most significant change related to this plan focused on the knowledge content. Students who took the integrated courses took an integrated test at the end of each semester that combined knowledge from different disciplines. Since in some cases different subjects refer to the same knowledge, the integration of courses reduced delivery of the same knowledge in different subjects and reduced the overlap between different disciplines. Meanwhile, for some subjects, the medical cases were also added to the curriculum to help students connect classroom theory to practical reality. The New Curriculum reorganized the knowledge content, yet the delivery method remained LBL. As one manager pointed out, the teaching method was "pouring knowledge into students' heads." In other words, this change was still "superficial" and did not touch the essence of the education— teaching and learning. Therefore, changing the teaching and learning method was still necessary for the staff and students at ZU.

Some individual teachers attempted to bring new elements into the traditional education activities such as case analysis in the Department of Social Science, and group discussions in the Department of Medical Ethics. Sometimes unmet practical needs spurred interest in these new elements. For instance, a teacher from the Department of Anatomy noticed the learning content was extremely boring for the students and learning was usually based on students' rote memory; to change this, she searched for a more engaging approach.

6.2 The Change Process

This section covers the process of the change towards PBL at XU. PBL was initiated by the managers, who established a funding project to stimulate the staff to participate. The management level did not initiate an institutional-wide PBL curriculum model; rather, the university encouraged the staff of each department to develop their own PBL approaches best suited to their own contexts. Therefore, the staff at ZU developed a variety of PBL approaches.

6.2.1 University Level

6.2.1.1 International Experience

The fundamental rationale for the change to PBL at ZU is described as "evidence based medical science," implying that any reform or innovation in medical science should be based on previous evidence. Following this, ZU searched for a peer experience from Harvard University in the US. Harvard University is a first class university that had undergone ten years of innovation from 1985 and 1995 and had accumulated rich experiences in educational reform. Their concept and practice of hybrid PBL seemed successful. Therefore, ZU sent a group of academic staff to participate in Harvard's workshops and seminars, to observe their lectures, classrooms and laboratories, as well as to discuss educational issues with their staff. The group also stayed at Washington University in Seattle for nine months to learn more about PBL. After they returned from the US, they asked the Head of the Department of Pharmacology of Harvard University to serve as a consultant and host a workshop about PBL at ZU. The American professor organized a 12-student group discussion, and then showed ZU staff members how to do so. Eventually, the top managers made the decision to introduce PBL to ZU.

Apart from this, the national Ministry of Education and the Ministry of Health were advocating PBL and integrated courses, providing more justification for the change to PBL.

6.2.1.2 Start-up Seminar

Prior to the formal introduction of PBL, ZU organized a series of seminars for staff members to mobilize them to participate in the change process. One of them involved over 200 staff members from both the basic medical and the clinical science fields. The main objective of the seminars was to inform staff members about PBL and to help them develop ideas about PBL implementation. The vice president for teaching and learning and other experienced staff members presented the relevant knowledge to the less experienced staff members who afterwards participated in a discussion. There were some doubts about PBL, which mainly concerned its feasibility in light of the lack of resources at ZU as well as students' and teachers' unfamiliarity.

6.2.1.3 Funding the Project

After the completion of these seminars, the vice president for teaching and learning launched a funding project, aimed at introducing PBL elements into the educational structure (Z University, 2004). The project was mainly administrated by the Research Center for Medical Education at ZU. The university encouraged all departments and staff members to apply for the funding project to enable participation in the curriculum innovation activities. Normally, the project required the cooperation of an entire department rather than just an individual. One teacher could only implement PBL in his or her own class but not in an entire course, making it necessary for other staff members in the same department to support each other. Therefore, in most cases it was the head of the department that led other staff members to apply for the project. After the approval of the PBL project from the university, the departmental head delegated various tasks such as searching literature, designing the problem (usually in the form of a medical case), and training the students, etc.

Each project was awarded 20,000 RMB which covered both the remuneration for teaching and other daily expenditures such as equipment (like the electronic white board for the Regional Anatomy course) and other official facilities. Since PBL called for much more staff effort in teaching, staff needed economic support to encourage their participation. There were human resource considerations, too. A big class lecture conventionally required two teaching hours. When the big class was divided into small groups, it required more staff members to complete the same task. Over half of the funding was spent on staff as an economic motivation. Nonetheless, the financial compensation was not as attractive as what staff could obtain by conducting research or doing clinical tasks.

6.2.1.4 Strategy: "All flowers blooming together"

The question of "What Is A Pure PBL Model" perplexed management at first. Soon, however, the managers realized that there was no fixed, standardized, single, uniform PBL to be found from international experiences or the literature. Therefore, the university adopted a flexible approach to introducing PBL, called the "all flowers blooming together" approach. This old Chinese saying indicates a flexible strategy of implementation. Top-level managers abandoned rigid design and allowed more freedom for the bottom-level practitioners to carry out their own exploration and innovation.

Thus, no systematic design or compulsory requirement was imposed by university management, so all departments across the university and its affiliated hospital had the freedom to decide their own concrete implementation strategy. This freedom allowed diversification of PBL across various departments. The rationale underlying this strategy was "to explore the suitable PBL model adaptive to Chinese context. The essence of PBL is student-centered learning, and promotes student's motivation. The rationale is more or less the same, but the method can be different," as one manager recalled. Basically, each department could determine:

- The proportion of PBL in a course (ranging from 0 to 100 percent);
- What content would be delivered in the form of PBL;
- The duration of PBL implementation (a few weeks to a whole semester);
- The classes or courses used to experiment with PBL (a few classes/all classes, integrated course/regular class, five-year program classes/seven-year program classes);
- How to design problems;
- The size of the groups;
- Parts of the assessment method.

The university received over 100 applications for the project and approved 60 of them at the initial stage in 2004. The change had passed three phases by the end of 2010, with each phase lasting two years. All the phases were carried out in the same manner: the university announced the funding projects, and the department or the staff applied for the project to introduce PBL elements into their courses. The first and second phase, from 2004 to 2008, was a "wide reform" period, meaning that PBL was implemented in a large number of courses. The third phase, from 2008 to 2010, focused on the reform of nine integrated courses. While the overall curriculum structure remained largely unchanged, the flexible strategy of implementing PBL allowed great diversity among different departments and courses. The Department of Physiopathology transformed all its courses into PBL modules, including four large cases interlinking all the basic knowledge. The Cell Biology course developed a clinical-case analysis model in the form of a small group discussion. The Medical Ethics course combined small group discussions on ethical issues in the student dormitory with group presentations in class. The General Surgical staff required students to participate in an online patient case discussion due to the lack of space and time to hold such a discussion in class.

There were sum-up meetings for PBL at the end of each year. The staff members, mainly the heads of the departments and the key teachers, met to communicate their experiences to each other. At the meeting, each project group reported on their experiences, problems and solutions and the other groups provided feedback. The university later gave an award to the project group who had excelled in PBL implementation, which motivated both the prize-winning groups and non-prizewinning groups.

6.2.2 Course Level

The university did not design a systematic curriculum approach to implement PBL; rather, it left the power of curriculum design to each department. The percentage of PBL in different courses varied according to the characteristic of each discipline, ranging from 100 percent PBL (e.g., Department of Physiopathology) to only a few hours (e.g., Department of Philosophy). The ratio of PBL to LBL components varied in different departments; for instance, in the Department of Bio Chemistry it was 1:2, in Cell Biology 1:9, in Pathogeny Biology 1:1, in Immunology 9:8, in Information Management 7:3, in Gynecology 3:7, Medical Ethics 5:2.

The majority of the courses employed a combination of LBL and PBL. For the course of Physiology, LBL served as an introduction to all the knowledge content that would be learned by students later through the PBL approach. Over ten chapters of the textbook (almost half of the course content) were selected for implementation in the form of PBL. Prior to the PBL unit, the teacher used two or three traditional lectures to rapidly give an overview of all the chapters that would be implemented in PBL and explain all the major points to the students. The students then learned these chapters in detail through a case analysis and group discussion PBL approach. This kind of

PBL can be viewed as preliminary lecture led PBL. In the Department of Bio Chemistry, there were three major content areas: structure, metabolization and inheritance. Metabolization was chosen to be delivered through PBL because it had a relatively fixed knowledge structure. Furthermore, presenting the second part rather than the first one with PBL offered students a "buffering phase" that allowed students to ease into the PBL environment. Both Physiology and Bio Chemistry used a partial PBL approach but in quite different manners. In Physiology, there was still a lecture unit as the introduction to the PBL unit, whereas in Bio Chemistry, the PBL unit was conducted in a pure manner without a supportive lecture. Therefore, the PBL approach in the course of Bio Chemistry can be seen as a partially pure PBL model in contrast to the 100 percent pure PBL model used in Pathology.

Medical Ethics implemented PBL in all of its seven large classes. The staff from this department used a traditional textbook, which could be divided into overview chapters and several "branch" chapters. The overview chapters (one third of the textbook) were delivered in lecture form, and the remaining chapters (the other two thirds of the textbook) were mainly delivered in the form of PBL. There were eight hours of lectures and 20 hours of PBL. The PBL part was composed of four medical cases, each of which consisted of four teaching hours: two for small group discussion in the student dormitory, and two for presentations in big classes. The case was closely linked to clinical medicine or major social events such as the relationship between doctors and patients, memory transplants, and the commercialization of human organs.

For an integrated course, PBL might take multiple forms in the same course because two groups of people were running different sections of the same course. For example, the course of Cell Chemistry and Biology was coordinated by the staff from the Department of Cell Biology and the Department of Biochemistry and Molecule Biology. Approximately two thirds of the course time was allotted to Biochemistry and Molecule Biology and the rest to Cell Biology. Separate departments running different parts of the same course lacked coordination. Different departments addressed PBL in their own way even within the same course.

The Problem unit in PBL at ZU often took the form of medical-case analysis. The length of the cases varied according to the discipline or necessary knowledge. The shortest case may only have lasted for one hour, while the longest required students to study for four or five hours in addition to the time needed to search and read the literature. Some courses assigned several intra-curricular hours for the students to search the literature relevant to case analysis, whereas others encouraged students to study the literature in extracurricular time and the intra-curricular hours were only devoted to group discussion.

As there were many parallel lectures for the same course and there were different classes (e.g. small classes, large classes), the department could select which parallel lectures they were going to change to PBL, and which class would apply PBL. For instance, the Department of Pathology selected a large class and a small class to implement PBL, the Department of Pathology chose to use PBL in one large class, and the College of Pharmacy implemented PBL into two large classes.

Interpretations of PBL varied. The majority of all departments employed the case study/analysis model. The P in PBL can be viewed as a medical problem. In the Department of Information Management, the P was a project (e.g., constructing a website).

Some changes occurred at the group level. Students tended to make a single, fixed and closed answer to the problem assigned to them in the early stage of group discussion. Gradually, students became ready to accept more flexible, diverse and open answers. Staff members noted that some students were quite reserved when expressing their ideas to others in the beginning; they behaved "politely and silently". However, they became more eager to communicate several classes later.

6.2.3 Assessment

The dominant assessment method remained the traditional paper-based test at the end of each semester. Tests included several different types of question such as fill in the blank, multiple choice, short answer, and short essays. Some staff members included medical case analysis questions. Generally speaking, students seemed to be more accustomed to the traditional assessment method. According to many staff members, the assessment method did not change much for two reasons. First, the performance of LBL students and PBL students required direct comparison. There were concerns that students might not acquire sufficient knowledge under the PBL curriculum. Therefore, the traditional assessment was needed to test student's knowledge acquisition after they finished the PBL course. Second, there were worries about the fairness of assessment methods. Some people regarded it as unfair to allow PBL students to be held to different assessment criteria since all students had to compete for scholarships, promotions and job opportunities after their graduation.

A few new assessment methods appeared at ZU such as group oral tests and academic papers. In the course of Regional Anatomy, the tutor developed a complex assessment approach consisting of quizzes, paper writing, and a final test. The quiz and the papers made up half of the final grade, and the final test made up the other half.

At the end of each class, there was a small quiz. The teacher selected a student from each group randomly and asked questions relevant to the learning content. Sometimes the teacher used the electronic white board for assistance. The academic performance of a student would serve as a reference to assess that of the other students in the same group. If a student passed the quiz that meant all the other members in the same group also passed it. This type of quiz pressured all the group members to work hard and participate in the group activities because they did not know who would be picked to take the quiz. Since the teacher had to face 30 students at the same time, she did not have sufficient energy and time to supervise each student, so she innovated this random selection approach to guarantee wide participation from the students in class discussion.

Paper writing served as a complementary but essential way to assess students' learning outcome. Students handed in a weekly mini-academic paper, typically a literature review, which was also part of their regular academic performance assessment. Since the students had limited time to participate in intra-curricular group activities, the teacher had to pressure students to learn actively. Writing papers occupied a large amount of students' time and energy. "Each student is required to complete a small paper based on their group discussions. The paper aims not only to address the questions proposed in the class, but also to formulate students' own concept." Since there were 13 lessons for this course, students had to finish 13 papers, a very time consuming academic activity. The tutor commented on this approach,

> "I let them analyze the anatomical structure of the human body after being shot and one student conducted a quite complicated analysis by considering the angle of being shot. Therefore, he made a complex ballistic trajectory study. It is quite impressive. You can see that students' thoughts are new and creative. "

The final exam in the Regional Anatomy course sometimes took the form of a medical case analysis. The teacher designed 11 points for students to analyze through discussion. The teacher simply showed the students the case without telling them the 11 "analysis points," And the students had to find the analysis points themselves. Normally, approximately half of the students successfully found all the 11 points that they needed to address. The "completeness" of covering all the analysis points stood at the center of the assessment, and the staff appreciated the students who were able to find all the analysis points.

6.2.4 PBL Procedures in Class

Except for a few departments, which adopted a full PBL model, the majority of departments had a lecture unit in which the teacher gave big lectures. Some departments only offered lectures about the non-PBL chapters of the textbook and left the students to learn the PBL parts through group work. Others rapidly covered the chapters or the components (in the form of lecturing) that would be delivered in PBL before the real PBL unit started. This section focuses on how PBL was generally carried out in the classes.

6.2.4.1 Task Assignment and Students' Training for PBL

The teacher assigned students a medical case to address at a later stage. Sometimes the teacher presented the full case to students, and sometimes the teacher showed the case to the students gradually, depending on the size of the problem and the teacher's curriculum arrangement.

Since most students lacked experience in self-learning, group discussion, and presentation, the teacher in most cases hosted a preliminary training program for students participating in PBL courses. The teacher showed the students what PBL was, how to practice PBL, how to organize group work, etc. Occasionally, the teacher hosted a small meeting with the class monitor or the group leader, informing them "the do's and don'ts of PBL" since student leaders play an important role in class affairs and group discussion.

6.2.4.2 Grouping

Students were grouped by different approaches: random grouping, student-ID-based grouping, individual-characteristics- (e.g. gender, social class, personality)-based grouping. At ZU, the size of a group was normally 10 to 12 students. The smallest group consisted of six and the largest consisted of over 20 students. Grouping processes were usually based on students' personal will, but in some cases the staff put great effort into grouping to yield a better outcome. In the Deparment of Gynaecology, the teacher grouped students on the basis of three indicators: student personality (introvert or extrovert), gender and social background (from city or country). The staff believed that by mixing students of different types together, students could compensate each other and thus make the group work more efficiently.

Normally, the size of a group was around 10 students. However, the ideal group size was six which was only possible in a few cases such as the course of physiology. Due to the lack of staff and physical space, some courses had quite large groups of approximately 30 students (e.g., some social science courses, Regional Anatomy). Facing 30 students at the same time was extremely challenging for teachers..

Some departments adopted a special way of grouping. They divided the class into six groups: one for presenting, one for recording the others' performances, and four "resting" groups with no official tasks. Students in this class took turns being members of the presenting group, the recording group and the resting group.

6.2.4.3 Group Leader

The leader of a group was usually selected from the students with better academic performances or with an active participation attitude. After grouping, the staff hosted short meetings for group leaders, informing them about the basic principle of PBL and how to implement PBL in each group. This was necessary because most Chinese students had no idea of what PBL was before they attended the PBL classes at ZU. The leader of the group received a small amount of preliminary training, usually taking the form of mini lectures hosted by staff. All group leaders might gather at the staff's office, where the teacher showed them the basic principles of PBL, how to implement PBL, what role the group leader should play, and what tasks the group leader should complete, among other things.

After this training, the group leaders returned to their own groups and assigned tasks to different members. Normally, a medical case was divided into several small units. Each member was responsible for one unit, searching for relevant literature individually through library and online resources. After the completion of this phase, the group members came together. The leader then collected the work from all of the group members, summed it up and returned it to the whole group in order to prepare for the group discussion.

6.2.4.4 Small-class Discussion

The large class of approximately 120 students was always divided into four small classes of 30 students for group discussion. Within each small class, there were three or four discussion groups. Beforehand, the teacher assigned the relevant medical case to students, and students tried to understand the case and solve the clinical problem through the use of multiple learning resources: library, internet, books, journals, etc.

For the small class discussion (see figure 6.2), the groups took turns directing and guiding the class activities. In the course of Physiology, each group was responsible for a medical case. When a group's turn came, that group assigned two chair-students to organize the class discussion. Students in the same group took turns being the chair students.

The small group discussion (see figure 6.3) was chaired by the group leader who took charge of the practical issues of the group such as information collection and task assigning, among other things. Sometimes a teacher or a higher-grade student was assigned to facilitate the group discussion or to record group members' performance.

The small class discussion usually took place in the following manner,

> "The physiology class is composed of 36 students, which is further divided into 3 groups, each of which consists of 12 students. The students in today's course are intended to learn the cardiac function through the exploration of a cardiac disease. Prior to class, a group is assigned to read through the case, search for relevant information, discuss and generalize their findings, and take charge of organizing today's learning activities. Today, this group assigned two male students to be responsible for organizing the group discussion. The group initially read the case to the cohort of the students, and afterward they posed some questions to other students. However, only a few students offered responses to the questions, and the majority of students kept silent due to either their lack of interest in the topic or the lack of preparation for the course. Therefore, the two students were forced to address the questions themselves and present the findings to the others. The teacher was present throughout the discussion, serving as an observer."

Figure 6.2 Small Class Discussion

Figure 6.3 Small-Group Discussion

6.2.4.5 Group Report or Presentation

If it was a large PBL class, after the completion of group discussion, students returned to the large class and one or two students from each group were delegated to communicate their findings and thoughts to the other students (see figure 6.4 and figure 6.5). The student responsible for this task made slides based on his or her group's results, stood on the platform, and presented to the other students. They presented the understanding of the case in their group, proposed questions and solutions to the clinical problem, and pointed out the connection between the case and their medical knowledge. The time allocated for each group presentation was around 15 minutes. In practice, it was much more like one student giving lectures to other students.

It should be noted that not all students had the opportunity to present to the larger group. The chance to present was dependent on the proportion of PBL in the course, the number of medical cases in the course and the individual personality of the student. The larger the proportion of PBL in a course, the better chance the students had to participate in the presentations. In the PBL courses that included several cases, the students in the same group might take turns presenting since students had to give a group report for each case. In this manner, the number of cases determined the number of students who could participate in the presentations. Also, extroverted students were much more likely to seize this opportunity than the introverted ones.

A typical large class group report often followed a similar pattern as described in below (see figure 6.4 and figure 6.5).

> "This is a big PBL class of Cell Biology. The size of the class is 120 students, further divided into 12 groups, each of which consists of 10 students. Prior to a class, each group is assigned a topic to research,

Figure 6.4 Students Making a Group Report 1 (from Biochemistry course)

Figure 6.5 Students Making a Group Report 2 (from Medical Ethics course)

discuss, and summarize before coming to class. This preparation is expected to be done in the students' free time. In class, the groups take turns introducing their topics and presenting their finding to others with the assistance of slides. Each group assigns one or two student representatives to give the presentation.

The classroom is quite large. Most of the time, the majority of the students who are listening do not respond to the students who are giving the presentations. Exactly the same things happen in these

> classes as the traditional lecture classes. Many students do not even listen to the presentations because they are distracted by working on other courses' homework or doing their own stuff.
>
> The teacher serves as an observer most of the time and replies very little to the presentations. It is only at the end of the class that the teacher makes a final comment."

The extent of the teacher's involvement in the PBL course varied and largely depended on the lecturer. In Pathogeny Biology, the lecturer was minimally involved in the group discussion, and he mainly served as an observer and a recorder of the students' performance. In Bio Chemistry, the teacher maintained that her responsibility was to guarantee that "the discussion is on the right track." Therefore, she did not interrupt unless the students started to discuss something irrelevant to the learning material. In Gynecology, the teacher was quite concerned with the knowledge coverage, and therefore, she was willing to challenge the students in the group discussion if she felt they had not covered the necessary material.

6.2.4.6 Finals

After the group report, sometimes the teacher commented on students' presentations and offered a conclusion or a summary of the course since students may have failed to capture the main points of the material by studying on their own. Since the Chinese education system was still focused on the retention of basic knowledge, the teacher at this stage summarized the main points to students because the students would need this knowledge for the final exam. This review gave students an idea of which points were useful and important for the assessment and which mistakes were easy to make. The staff also, in some cases, gave students feedback on their performance in the course. One teacher even made some small awards for the student who performed best in the PBL course.

Some staff members wanted to know students' attitude towards their teaching and the PBL course. They conducted in-class surveys in order to investigate student's attitudes towards PBL. Based on these findings, staff could make some adjustments at a later stage to better meet students' needs. Other staff members used alternative means of gathering feedback. For instance, after the completion of the Medical Ethics course, the group leaders were invited to an informal staff-hosted party aimed at communicating their feelings and

Figure 6.6 Student Discussing Problems in Their Dormitory

reflections about the PBL course to the teachers and proposing ways to improve it. The group leaders came up with a series of recommendations such as abolishing instructions for discussion because it restrained students' thoughts.

6.2.5 Dormitory PBL

Normally, PBL takes place in the class. However, due to time and physical space limitations at ZU, the staff innovated several new ways of practicing PBL in order to overcome these constraints such as dormitory PBL and online PBL.

Dormitory PBL (see Figure 6.6) was an answer to the problem of limited space at ZU. One obstacle to implement PBL at ZU was the shortage of classrooms for small group discussions. Since a large number of undergraduate students lived on campus, some staff members used student dormitories as a place for group work. When it was time for small group discussions, students used their dormitories to hold this group activity. Often, they selected geographically neighboring rooms for convenience. During the discussions, teachers came around and supervise them as well.

Although dormitory PBL was a formal part of courses, group work was held in the dormitory only because there were not enough classrooms for students to use for group activities. Students' performances in dormitory discussions were taken into consideration for the final assessment, even though there remained a sense that the dormitory was an informal place for learning.

The use of dormitories in implementing PBL group work appeared in many courses, which were in need of classrooms. The researcher observed a typical dormitory-based PBL class as following:

> "It is a group discussion from the Medical Ethics course that takes place in the students' on-campus dormitories. Each dormitory consists of a medium-size group of ten students. The supervisor, since she supervises two groups at the same time, she has to cross supervise two dormitories and leave the responsibility of organizing discussion activities to the group leaders. When she is together with the students, she mostly appears as an observer and sometimes participates in the group discussion.
>
> The discussion is concerned with an organ transplant case and the discussion is extended to debate whether a transplant organ can be merchandized or not. The aim is neither to write a report nor to seek the solution of the problem, but to discuss and understand ethical issues arising in this case. Since the students received the reading material a week ago and are expected to have read it, the group directly jumps to the discussion without reading through the case description again. It seems that the students are quite interested in this topic and great group dynamics can be seen in the group. Some students argue for the necessity of merchandizing organ transplants due to the lack of organs, and they draw upon numerous examples from their friends and the news. Others insist on a conservative approach to dealing with this matter. The group leader is both an organizer of the group discussion and an active participant in the discussion.
>
> There are two students who are relatively silent. They seldom speak."

6.2.6 Online PBL

The Department of Pathology and the Department of General Surgery developed a PBL approach, specifically an online PBL model utilizing information technology. Online PBL served a "complementary role" for the traditional lectures and intra-curricular activities due to the limitation of intra-curricular time and physical space. There were two types of online PBL at ZU. The concept of online PBL appeared as a replacement for the group activities normally used in PBL courses. Partly, online PBL was inspired by Japanese experiences. In the General Surgery course, the teacher gave regular lectures and assigned students medical questions and cases. There were very limited group activities and limited teacher-student interaction in the course since the teacher had to finish his teaching in the intra-curricular time. After the course, students addressed the questions and cases independently through informal discussion or searching online resources. They could turn to the online PBL BBS for help with problems. The BBS was called Ding Xiang Yuan, which was one of the biggest medical professional BBS in China. The teacher created a special message board for the course and he assigned each student an ID. When students had problems regarding their course and the relevant medical cases, they could log into the BBS with their ID, and publish questions in the message board to the teacher, who would answer them online. In this way, the online BBS system allowed for improved interaction between the teacher and student. Since the teacher positioned this online activity complementary to the course, he did not assess students' online performance; neither did he integrate it into the final assessment. However, there was a compulsory requirement that all students participate in the online activities. Since then, the Department of General Surgery completed their PBL project and has discontinued this online BBS system.

6.2.7 Staff Training

There were two types of staff training at ZU. The first was general staff training, which the university required of teachers just beginning their careers. New staff had to develop their teaching plans with experienced staff, audit old staff's lectures, and give trial lectures to other teachers in the same department. In doing these activities, they would become qualified university teachers by communicating with their colleagues.

The second type of training was specifically for PBL. Prior to the implementation of PBL in 2004 at ZU, the university organized a series of meetings and workshops to inform staff members about PBL and how to implement it. However, it was only available to key managers, departmental heads, and some key teachers. The majority of the remaining staff only received the "second hand" training delivered by the key managers, departmental heads, and the key teachers who are accessible to the first-hand PBL materials and experiences. A PBL start-up meeting was open to almost all staff members, as was an annual sum-up meeting at the end of each semester.

6.3 Issues and Challenges

The change towards PBL was not easy; numerous and diverse issues and challenges arose during PBL implementation.

6.3.1 Teachers' Role

There were significant questions about the teachers' role in PBL. These included: how should teachers behave when they served as facilitators in the groups? To what extent should the teachers participate in the student's group process? What the strengths and costs when the teachers give students sufficient freedom to learn on their own? At ZU, the teachers' behavior varied depending on the department. In the course of Regional Anatomy, teachers were actively involved in the group discussions. In some departments, teachers only served as observers, recording all students' performance, such as the frequency of their attendance, the quality of their proposed questions, the duration of their interaction with other students, which partly served as the basis for the final examination. They did not interfere with student discussion until students made mistakes or veered off topic.

The issue of the teachers' role was closely linked to the resource problem. The majority of the interviewees complained about the lack of teachers. They argued that PBL was staff-intensive. For instance, a normal large class at ZU consisted of approximately 120 students and required one teacher, but the same class dividing into 12 groups required 12 teachers. Therefore, the PBL method required more staff members than the traditional teaching approach. Some teachers regarded the lack of staff as an insoluble problem, whereas others believed that this problem could be easily resolved as long as the teachers served merely as observers and the university could utilize graduates or higher-grade students for class assistance.

6.3.2 PBL and its Value

Understanding of the P in PBL varied across different disciplines and staff members. The dominant form of P was a medical case, although there was ongoing controversy over the size of P, its position in the course, and its percentage in the whole curriculum system. One teacher defined PBL as "through the use of clinical examples, inducing students to integrate the knowledge they learn in the classroom, inspiring them to develop their critical-thinking, problem-finding and solving skills, rather than directly showing them the textbook." Another young teacher viewed PBL as a way "to motivate students' interest in learning, engage them in class and cultivate their group ability, etc."

In a few instances, P took the form of a mini product, such as in the disciplines of Bio Engineering and Information Management. The social science department treated "P" in a much broader sense such as reading classical works or studying case-based material from science philosophy, or history. One interviewee from the social science department proposed three kinds of cases/problems: the classical work of social scientists, the history of scientific history, and the philosophy history. In practice, the teacher selected several chapters or paragraphs from the classical work and assigned them to the students to read. Sometimes, a case in science and philosophy was also used to assist students in learning and understanding. After reading the material, the students discussed the case or the reading material in small classes or groups.

Still, some teachers argued that LBL was more advantageous than PBL in many aspects. The justification was that sometimes lectures were more efficient than PBL. A staff member commented,

> "For some bio chemistry knowledge, teachers can make it clear and explicit to students in a half an hour lecture. However, students probably remain confused after one morning's self-learning if they learn on their own. In this sense, PBL is not very efficient and you can see that some knowledge is not suitable to be delivered through PBL."
>
> Another senior staff concluded,
>
> "Some knowledge is not suitable to be delivered through PBL. The 'trunk' content should be taught in lectures, and the 'branch' knowledge can be used to motivate active thinking. If you deliver the 'trunk' knowledge through PBL rather than lectures, students will lose themselves. . . . the 'trunk' must be taught in the form of lecture." "It is not going to work if you start the course with PBL rather than a lecture."

For basic medical courses, there was a commonly agreed notion that PBL should be centered on basic knowledge, especially from the textbook. "The medical course is important, yet the basic knowledge is even more important." Thus, the small group discussion should focus on fundamental knowledge rather than the medical patient, although the patient case may be intriguing. The staff insisted that it is important to "grasp the basic knowledge" "formulate a relatively well-rounded knowledge system" and "consolidate students' knowledge base." There were two justifications for this argument: first, the final test covers mostly basic knowledge; second, "only those students who have a solid knowledge base can become qualified doctors after their graduation." Some managers also highlighted the importance of knowledge retention for students' future development: "the curriculum is fixed, if students don't learn this stuff, what they can do after their graduation? Who dares to trust a doctor without a consolidated knowledge base?"

The teacher's direct role in the teaching and learning process was also emphasized by some interviewees, as they commented,

> "Teachers make a difference in student learning. You should know that, even a simple lecture is based upon a teacher's knowledge and experience which has been accumulated over many years. A lecture represents a teacher's effort in preparing a 45 minutes lecture. It is not so easy. We can teach surgery because we have many years' experience operating. Students' activities such as searching online resources are incomparable with the expertise of the staff."

6.3.3 Attitudes of the Staff

PBL had not gained wide recognition from staff members throughout the university. It placed a burden on staff members. PBL requires more significant time and energy preparing for the courses and supervising the students than traditional lectures in which teachers only needed to present knowledge to students. In addition, PBL only involves a particular group of teachers who were interested in improving teaching quality. Staff members who did not participate in PBL projects did not pay much attention to the teaching issues. Even staff members who joined the PBL project sometimes treated PBL as a task imposed by the PBL funding program, rather than reflecting their conception of teaching. This meant that after they "finished" the PBL task, they returned to the old way of teaching.

Several practical issues affected the implementation of PBL. Firstly, staff members at the university had many tasks related to clinical and research missions. The teachers at the affiliated hospitals of ZU were overwhelmed by clinical missions, such as surgery, and thus they did not have sufficient time to teach. Secondly, doctors treated teaching as only a minor portion of their jobs, thus they were not eager to experiment with PBL. Thirdly, the staff gained many more benefits from conducting research than teaching due to the university's emphasis on scientific research. Income and promotion depend largely on achievements in research rather than teaching. Therefore, the majority of the staff members focus more on researching than on teaching and were not very interested in spending time implementing PBL. Finally, teaching quality was seldom linked to staff's actual income since staff earned the same regardless of their teaching quality. This factor convinced the staff that,

> "Good teaching is the same as bad teaching. . .it is enough to fulfill the teaching obligation. . . it is good enough as long as there is no mistake in teaching."

Some staff members insisted that "we are aimed at producing generalist. Therefore, the students have to grasp the whole structure of the medical knowledge system. They have to learn." The staff members at ZU were generally concerned for their students.

A senior staff member referred to the specialty of the medical field as he commented:

> "The medical domain is special. Medical knowledge content cannot be fully understood if the students only learn on their own. Sometimes students may get stuck in the learning process; in this case, a slight guidance from the teacher can make a big difference."

Staff also expressed concern in extending PBL to all undergraduate students because the "the good students and the bad students are intermingled," and "students lack the ability to learn by themselves." Since the students lack experience in a PBL environment, sometimes staff needed to give students instructions. For instance, students might not know where to start when they begin to analyze a medical case. Therefore, staff offered them some clues to facilitate a group discussion. However, as long as staff members did so, students were probably constrained by the staff's instructions and failed to explore on their own.

Another teacher with a background in ethics insisted that the students were unable to learn ethics well if they were not equipped with some preliminary knowledge about the subject. She noticed that students who discussed medical ethics mostly used "daily language" rather than "professional language."

The teacher supposed that because the students did not have a philosophical knowledge foundation, they could not discuss ethics at a professional level. Apart from this, the head of the department highlighted the significance of the fundamental theory and knowledge students needed, and he, therefore, attempted to increase the lectures rather than reduce them.

There was a general recognition that Chinese students would not adapt well to a pure PBL model. As a young staff member concluded,

> "I think we could mix pure lecture and PBL, but we should give the freedom to teachers to decide the way of mixing them, the portion of PBL and lecture, etc. You cannot use the 'either white or black' thinking."

6.3.4 Students

Implementing PBL was also quite a challenge for students.

> "It is not going to work if we don't motivate students' interest to learn even though we have good quality teachers. On the other hand, it won't work either if students are not eager to learn even though our staff works hard."

The managers claimed that the change toward PBL gained wide recognition and support from students compared to other educational innovations at ZU. "They recognize that they grasp the knowledge by learning rather than memorizing. This time, from PBL, they know how to learn."

Some students were eager to see an increase in PBL learning hours throughout the whole curriculum. They recognized the benefits of PBL, as shown below in a quote from an interview:

> "It is better to study in the PBL model than in traditional courses in that the retention of the medical case, the symptoms, the solutions are much better because I learn them by myself."

> "I forget all the knowledge immediately after the final test at the end of each semester. But if it is learning through PBL, there is more retention in my head because I learn it by myself."

Students found PBL "interesting" or, "good." They sometimes experienced "fierce" discussions in the group activities. Even after class, they still continued to communicate their thoughts to one another. Presenting in front of the large class was a valuable chance to overcome anxiety. "You have to overcome the fear of being in front of a large audience. You have to make your presentation attractive and do many things." Many students recognized the benefits of PBL:

> "I like PBL. I think the content in the textbook is constrained. We can find much more than the textbook tells us. When I made the slides for a presentation, I can feel it. Since everyone participates in the group discussion, we have a better impression and remember more of what we have learned."

> "PBL is intuitively different from traditional learning. If you use PBL as a teaching method, we can memorize the knowledge better. It is good way to communicate and understand the material being taught and to know how it is applied in real life. If we have questions, we can search the internet afterwards and communicate our findings in the class again. Sometimes, the teacher also helps us. It surely leads to a better learning result."

> "What we learn from the PBL course is quite complicated. First you searching the literature and you have to prepare a lot for the presentation. After the discussion, you understand the medical case better, especially better than you would have from learning by yourself. You learn the opinion from other people and the whole approach facilitates your later memorization."

The PBL project at ZU also imposed heavy academic burdens on the students. Since there was no overall design of implementing PBL, the university proposed the "All flowers blooming together" plan, which allowed each department to carry out PBL in its own way. Thus, the departments had the freedom to decide when to implement PBL, how to deliver PBL, and how to assess students. The problem was that sometimes there was no PBL whereas at other times the students would become overwhelmed with several PBL courses at once. For example, some students felt that they "collapsed" when they had to prepare for four PBL courses in one week. In some cases, students had over 40 hours of courses per week, and they were overextended and mentally exhausted by their courses:

> "Currently we have two PBL courses per week. Our normal workload has us in class all day long. Apart from this, we have to use our

> extracurricular time to prepare for group work, while still participating in our daily student community life, doing literature searches, and struggling to do well on the English Band 6 Test. . .we are so tired."

Some students saw their undergraduate life as an extension of their high school life. "Being in ZU is like in Grade 4 in high school."[1] The curriculum schedule is completely full, and they have to complete nine courses in one semester, leaving them with little spare time. Usually, they take medical courses on weekdays, and they attend elective courses in the evening or on weekends. After they finish with intra-curricular tasks, they have to do their homework. Therefore, students at ZU are quite busy.

Sometimes, there were four courses implementing PBL at the same time and preparing for PBL took a large amount of time. This was especially true when the medical problem contained terminology or concepts that students had not learned. Students had to understand the unfamiliar terms first in order to analyze the medical case. They had to spend excessive time and energy reading books, searching relevant literature and reflecting. The ratio of preparation hours for PBL courses to teaching and learning hours for PBL courses were usually 1:1, sometimes 3:1, or even 4:1. This implied that the students sometimes spent four hours reading through the literature before they had one hour of discussion. As one active student recalled, they sometimes had to prepare five hours for one hour of discussion in class.

> "The biggest problem of PBL is that it occupies too much time. The teacher left all the task of searching literature to extracurricular time. Whenever there is a need to search the literature, the teacher always told us to do it after class. However, if the teacher could leave one intra-curricular hour for us to search the literature, and then continue the intra-curricular discussion, we would be better prepared for it. We would also be able to digest the knowledge better. If the teacher lectured a bit slower, I would understand the material better."

For some students, PBL was a waste of time because it did not deliver as much knowledge as traditional lectures did in the same time. Some students

---

[1] In China, there are three grades in high school where students have to study extremely hard to compete for limited university entry opportunities. In some cases, students have to work over 14 hours per day during all 3 grades in high school, especially in Grade 3, the final grade of high school. The statement that university life is like Grade 4 in high school indicates the magnitude of students' academic burden.

stubbornly preferred traditional teaching and learning such as listening to lectures, memorizing the knowledge, individual learning, and they did not like PBL. Ayoung teacher commented,

> "They hope the teacher will tell them the major points after the lecture, so they can memorize those points and, practice. A lot of students do this."

A great many students also highlighted the need for knowledge acquisition in a PBL context:

> "My feeling is that in order to implement PBL, some teachers have to 'rush through' the lecture to 'squeeze in' the time for PBL. They lecture much faster than they do in non PBL courses. However, this 'rushing' makes me feel that I can't consolidate my knowledge."

> "We don't mean that we can learn a lot from one single course. I have a sense that the logic in the PBL course is not good. The course is not clear enough; it is a little bit messy."

> "It would be nice if teachers could give us some instruction or lectures about the chapters that will be delivered in the form of PBL. If we start with a PBL discussion, we are afraid that we cannot do it well because we lack the basic knowledge."

Some students who finished their PBL courses even went to the traditional lectures since they were not confident in the knowledge covered by PBL.[2]

> "It is difficult for us to understand the medical cases assigned to us because of the lack of clinical science knowledge. Somctimes we search the internet for help, but we still do not understand them... We have not learned pathology, so we have no idea of the medical examination result. Those English symbols, abbreviations, medical criterion and indicators, we don't know them at all....It is better to implement PBL after we learn all the fundamental knowledge, or after we step into the higher-level classes or the clinical phase."

[2]A further clarification must be made here. There are many parallel courses for a particular course due to the large number of the students. For example, a basic medical course is mandatory for the entire 1,200 undergraduate student body. As the undergraduate students may be further divided into 10 large classes or 20 small classes, one particular course may have 10 or 20 sections going on at the same time.. Basically, they cover the same learning material. Therefore, students have the chance to listen to the same presentation several times because they can attend different parallel courses.

Some students attributed their emphasis on knowledge acquisition to their previous educational experience, which shaped their learning habits as well as their attitudes toward PBL and the quality of PBL courses. Primary and secondary school education was lecture based and examination oriented. The main form of education was teachers presenting knowledge through lectures and explaining the essential fundamentals and difficult points of the material. This was followed by students memorizing these points and repeating them at the exam. Teachers controlled the whole learning process, and students had little freedom to learn by themselves. Learning was an individual process rather than a group activity. The main objective, for most students, was to obtain a high score on the examination. Gradually, students developed the habit of passive learning. They relied upon their teacher to tell them what to think rather than embarking on a discovery process of their own. Some students justified lectures by arguing that "lectures are more explicit, logical, and systematic in presenting knowledge" than PBL courses were. They had a sense of "chaos" in the PBL environment. Another myth students often perpetuated was the need for consolidating the knowledge basis, which they believed could only be achieved through lecture-based learning. In some extreme cases, students who attended PBL courses used their free time to attend lectures of the same course that were not using PBL in order to acquire more fundamental knowledge.

Other students became reflective of their own educational experience,

> "Although basic knowledge is quite essential for obtaining the academic degree, ur, how about the educational quality? What does it mean for students? If you ask me questions concerning the content I learned last month or last semester, I have totally forgotten it."

Some students also linked the issue of PBL to their overall educational objective:

> "If the education is only for the knowledge base, then the traditional ways of delivering is perfect. Teachers lecture and we memorize."

> "It makes sense to require students to grasp intra-curricular knowledge. But if we want to meet the needs of society, we are useless using this kind of education."

Some students reflected on their primary and secondary school experience, which they considered responsible for their lack of motivation to learn at the university:

> "We receive traditional lectures from our primary school to high school. There have been a huge number of education innovations, but our education did not change very much. Teachers still lecture and students still memorize and take tests." "After we get accustomed to this type of education, we become confused when we bump into PBL. From secondary school on, we feel a lot of pressure and form the habit of learning when teachers are watching us and slack when they are not. If we become more self-motivated in learning, it is unnecessary for teachers to superintend us."

At Chinese universities, students do not only take courses from their own major, but must also fulfill the requirements of the national curriculum, often called "general education curriculum" which includes foreign language, computer science, moral and political education and physical education. The general education curriculum constitutes roughly 30 percent of the whole curriculum. Moreover, in order to obtain a satisfactory job after their graduation, the majority of Chinese university students must obtain proficiency certificates (e.g., Chinese English Test Band 4 and 6, National Computer Test, etc.), participate in internships off campus, and prepare for National Entrance Examination for Graduates. These activities take tremendous amounts of time and energy from the students and take time away from them learning their own disciplines. The higher the class is, the less time students can spend studying their own disciplines. This universal phenomenon in Chinese universities has already posed great challenges to specialized education. In the case of ZU, this system significantly limited the time and energy students could dedicate to preparing for PBL courses.

The enthusiasm of students in PBL courses decreased overtime. In the early stage of PBL, students were quite active in participating in class discussion on problems. However, as time went on, the number of active students reduced significantly. They found PBL "meaningless and not interesting." "I want to say something about the case in the class. But after I hear other students' statements, I become suspicious that my opinion is probably be wrong." The fear of making mistakes bothered some students. "The discussion on the clinical case is too far away from the knowledge in the textbook, and it has little to do with the final test." "There are too many professional terminologies that we are not familiar with, so we think it is boring."

6.3.5 Resources

Resource unavailability was the issue most discussed in interviews. Managers, staff members and students regarded the lack of resources, mainly in terms of small classrooms, inadequate staff numbers, and limited access to libraries and the internet, as a significant challenge. PBL favors small group discussion and therefore requires a large number of staff members and small rooms. However, due to the resource reality at ZU, which includes a limited infrastructure and continuous staff shortages, the necessary resources could not be easily acquired in a short time. In the Department of Medical Ethics, for example, there were only three staff members to give lectures to seven large classes containing a total of 1,300 students. As a teacher commented:

> "The teacher student ratio at Harvard and Washington University is almost 1:1, but for us it is 1:16." "The key reason for good quality PBL at Western universities lies in their small number of students. We have too many students, over ten thousand per year. Western universities do not have so many students. The size of our graduate class is more or less the same as their regular undergraduate class. We recruit around 800 PhDs each year, plus master student, totaling over 1000 graduate students per year."

Normally, a teacher had to teach 120 students in a large class and 36 students in a small class. After placing students in groups, even relatively large groups containing 10 or 12 students, the staff needed to supervise and facilitate the groups multiplied exponentially.

In order to overcome the shortage of classrooms, ZU employed dormitories, staff meeting rooms, and online resources to compensate for the lack of space for group discussions. However, this was still not enough. Sometimes, students had to use a large classroom for small group discussions, where different groups sat in different corners of the same room and conducted their group activity. In an extreme case when there were not enough desks, students had to use a table tennis table as an alternative workspace.

In addition to difficulty finding adequate space, a great many students were also frustrated by limited access to the library and e-resources. They are not allowed to connect to online library resources from outside the library, forcing them to go to the library, a place which is often overcrowded with students. Sometimes, it was even impossible to find a seat to sit down and study at the campus library. The hours of library operation are also quite limited. Furthermore, students from the five-year program live off campus, making it even more difficult to access and to search online library resources. Sometimes these resource restrictions had a negative effect on their performance in group discussions in class. Although ZU was making efforts to improve its infrastructure, and the staff was recruiting an increasing number of graduate students as facilitators for group work, the lack of resources remained an unsolvable problem.

6.3.6 Assessment

The dominant assessment method for students remains the traditional paper-based test, consisting of several types of questions (e.g., fill in the blank, multiple choice questions, short answer, vocabulary, short essay) even five years after the introduction of PBL. Open-book testing was not allowed according to official university regulations. In order to make students work hard, the university mandates that all departments have formal closed-book examinations with fixed answers mostly concentrating on the knowledge in the text book. Students are not allowed to consult any reference when taking the tests; they can only use their memory.

Many staff members maintained that there were justifications for retaining traditional testing methods. First, traditional tests remained the most feasible way to assess students' academic performance before the invention of a new assessment method suitable for PBL. Second, the adoption of a uniform examination allowed academic performance comparison between PBL students and non-PBL students. If PBL classes and non-PBL classes employed different assessment methods, it would have been difficult to compare the PBL outcome and the LBL outcome. Therefore, the existence of traditional tests became an opportunity to examine PBL effectiveness. Finally, the students at Chinese universities experience ongoing competition for academic awards, supervisors and job opportunities. For instance, their opportunity to find a job at the university is based on their performance responding to 200 multiple-choice questions based on memorizing basic knowledge. In the seven-year program, students who earn the highest scores on tests are given priority in choosing their master supervisors. When it comes time to choose clinical disciplines (e.g., surgery, internal medicine, gynecology, pediatrics, etc.), test scores also determine the students' priority. Since the criterion for competition is mostly students' score on the final course exam, it is necessary to have a common platform to show the results of students' academic achievements.

In other words, a standardized-traditional test makes it possible to compare different students' academic achievements. If there were multiple methods to assess different students' learning outcome, it could cause confusion in comparing different students' academic achievement and be unfair, at least to some students.

6.3.7 Other Issues

The quality of integrated courses remains a major issue in ZU. Newly developed integrated textbooks, group teaching involving staff members from different disciplines, the Big Integrated Class 1 and integrated assessments were inherited from the integrated course project experiment. There was no doubt that the spirit of discipline integration had been embodied by educational practice. Nevertheless, from some interviewees' perspective, integration was nothing but "a large discipline superficially blending various disciplines." The content of integrated textbooks could be easily recognized as a combination of several isolated disciplines even though the title pointed to integrated material. Staff only took responsibility for their own disciplines rather than coordinating with each other, and the final exam was also divided into different disciplines (e.g., if the total score of a test was 100, 30% came from one discipline and 70% came from another). Sometimes the integrated courses still used several isolated textbooks. For instance, the course of Cell Biology and Chemistry had two textbooks: Cell Biology and Biochemistry. When it was a class on cell biology, students were expected to bring their cell biology textbook, and when biochemistry's staff members gave lectures, students took out their biochemistry textbook. The problems associated with integrating courses posed significant challenges to PBL implementation since PBL encourages interdisciplinary learning.

Furthermore, there were arguments about the feasibility of implementing PBL in different disciplines. Pathology and Pathophysiology were generally regarded as the most suitable discipline for implementing PBL since clinical cases were their main learning content. However, some basic courses which involved a large amount of factual information, such as mathematics, physics, anatomy and cell biology, presented more of a challenge for PBL implementation. Some staff members were a bit "reluctant" to implement PBL in the basic courses. However, when considering the working traits, staff at the affiliated hospitals seemed unable to conduct PBL since they were busy fulfilling clinical obligations. On the other hand, the staff members from basic courses were more likely to have the time and energy to carry out PBL activities, as some hospital staff insisted.

Finally, some problems regarding the micro-system design were noted by the staff. The limited number of PBL hours made the staff members uncomfortable in some departments. For instance, the Regional Anatomy course only had four PBL courses, and since students had little PBL experience before entering the university, they needed time to adapt to this new learning environment. For most students, the first and second exposure to PBL was their "adaptation phase" where they learned what PBL was and what to expect from PBL. It was only the third and fourth class that students began to "finish their adaptation phase" and "get into the groove." However, after the fourth PBL course, PBL finished and the traditional lectures returned.

6.4 Current Situation of PBL at ZU

The change to PBL at ZU had not yet been completed, but due to management changes and the completion of the PBL funding project, the process of change slowed down. After three phases of change, some departments abandoned PBL and returned to traditional lectures again. The Department of Bio Chemistry and Molecule Biology did not apply for the third project phase. They replaced the small class group activities with large class discussions. They abandoned medical cases analysis and adopted some intra-curricular questions instead. At first, the Department of Topographic Anatomy abolished all large lectures and replaced them with medical cases students were expected to analyze. After a while, however, the staff recognized that students in this system failed to "grasp the main points of the knowledge" and "formulate the theoretical system in mind." Therefore, they reinstated the large lectures to better equip students with fundamental knowledge. Physiopathology wished to reduce the portion of PBL in their courses because they felt that a 100 percent PBL course made students feel insecure about acquiring the covered knowledge. The Department of Pathology, which is quite famous for its online PBL model, finally abandoned online PBL since some staff members maintained that "the quality of an online PBL approach is weak in promoting the communication among the students."

Some basic medical courses and almost all clinical courses discontinued PBL after completion of their PBL projects (e.g., Department of Bio Chemistry and Molecule Biology, Department of General Surgery, Department of Gynecology, Department of Philosophy). For them, implementing PBL was no more than a task imposed on them by upper management. After they successfully finished their administrative duties, they were not willing to do it anymore. Small classes and group work left and large teacher-centric lecture classes returned. Meanwhile, a few departments continued their PBL innovation due to their own interest and ambition, such as the Department of Regional Anatomy.

As a long-term strategy, Big Integrated Class 1 in both the five-year system and the seven-year system had officially been designated as a PBL experimenting class where some classes must be taught using PBL.

Even courses that stuck to PBL reverted back to large class discussions instead of small class discussions. Though economically convenient, this drastically reduced students' opportunity to participate in class discussions. Students had much more opportunities to present, communicate, and question in small classes. In large classes, only a few students can communicate their findings and thoughts with others. In large group classes, most of the members did nothing but search the literature and prepare information for one student who presented on behalf of the group. In classes, most students only listened.

Some staff members commented that in general "PBL at ZU is declining now." Though it may sound frustrating, some staff members contended that it was too early to assert that nothing had changed because "there are some minor changes deep in people's mind" "they have already tried something new." Some departments are still using PBL; for example, the Department of Bio Chemistry still uses small class discussions and large class heuristic teaching methods.

Furthermore, PBL implementation has already made a substantial impact on the whole university.

Firstly, PBL practices have been partly institutionalized into the university system. For instance, the Big Integrated Class 1 in both the five-year system and the seven-year system being designated as the long term PBL experimenting class. Also, some PBL textbooks have been published to replace the traditional ones and have been popularized throughout the university. These PBL textbooks are the first ones in China and have become National Class Textbooks. Furthermore, teaching hours for PBL eventually gained the official recognition of the university, meaning that staff members can now gain more economic support for their participation in PBL teaching activities. Although the allowance for PBL teaching is less than the allowance for conducting research, it is nonetheless a good start. All these signs indicate that PBL has become part of the university routine.

In addition, the concept of PBL is now known to nearly all university staff members. The introduction of PBL is first and foremost a conceptual change project. Prior to PBL introduction, students and staff members were quite accustomed to carrying out their educational activities in a traditional fashion: teachers present the knowledge and students receive it and reproduce it at the final examination. After the PBL introduction project, staff members and students began to realize that there are alternative ways of practicing education, whether they like or dislike it.

Finally, the PBL project at ZU also garnered wide recognition from the outside world. Many courses that incorporated PBL won the National Quality Curriculum award. The achievements of implementing PBL at ZU drew the attention of peer universities from mainland China, Hong Kong, Taiwan, the US and other nations.

After managerial changes, specifically the retirement of the former vice president for teaching and learning, the commitment to PBL became less strong at ZU. On a positive note, although the scale of implementing PBL is smaller, the quality is constantly improving. Previously, PBL was conducted in a large number of courses, whereas now it is used in the courses that are most suitable for PBL. PBL will continue at ZU; a senior manager explained the necessity to maintain PBL at ZU:

> "PBL is a bright spot for ZU across all medical universities nationwide. It distinguishes ZU from other universities, and therefore, it must be maintained."

7

Interpretations of PBL Implementation

This book is concerned with the challenges and obstacles for higher education institutions when they make the educational transformation to a PBL system. As noted in the first chapter, the literature lacks studies that employ a systematic perspective and studies that address organizational members' subjective worldview when dealing with the challenges and obstacles for PBL implementation. Therefore, this chapter will address how the organizational members understand and interpret PBL implementation at XU and ZU. Primarily, this section is concerned with three major themes interpreted in various ways by the organizational members, namely: PBL in general, social learning in PBL and motivation. Since PBL is the key word in the educational change at both universities, individuals' understanding of PBL is crucial. Managers' and staff members' conceptions of PBL largely influence how PBL is actually delivered in practice. Social learning is another essential dimension of PBL principles (Kolmos, 2008). Compared to the interpretation of PBL in general, which concentrates on curriculum design, or the macro level, the examination of social learning helps us see how PBL works at the micro level. The motivation of the organizational members is a key factor in sustaining change including the change to PBL (e.g., Kotter, 1995; Moesby, 2004; Cohen, 2005). It is therefore necessary to identify what resources at the university can motivate the organizational members to engage in PBL activities, and what factors will demotivate the staff members and students from participating in PBL activities.

7.1 Different Interpretations of PBL

PBL theory includes various types of PBL definitions and practices (Barrows, 1986; Savin-Baden, 2003; Savin-Baden & Major, 2004). In addition to these, some notions, such as project-based learning and case studies, have also become relevant to, or even synonymous with PBL (Boud, 1985; Barrows, 1986; Savin-Baden, 2003). Each of these definitions, practices and ideas can be perceived as a particular way of interpreting PBL and representing educational values. It is generally acknowledged that PBL cannot be viewed as a particular type of educational approach; rather, it should be conceived of as a philosophy covering many variations and forms (Savin-Baden & Major, 2004). It is worth noting that at XU and ZU there was no consensus on a single interpretation of PBL; rather, a diversity of PBL interpretations emerged, which had a strong impact on the organizational members' practices. This section will specify how PBL is understood and interpreted in practice by the organization as well as by its individual members, and how that understanding influences the way PBL is put into practice.

7.1.1 Different Interpretations of PBL at XU

At XU, although the university attempted to introduce PBL at the organizational level and impose it throughout the entire organization, many variations of PBL interpretations can be observed. These will be laid out in the following section.

7.1.1.1 Different interpretations of PBL at XU 1: Medical PBL

Some staff members, due to their own work experience, regarded PBL as being the same as the methodology widely used in the medical field, in which students work together on a medical case. A senior staff member with many years' experience of staff training, researching tertiary education, recalled:

> "I worked in the medical field before I came here. That was where I was introduced to PBL. But I chose to use PBL when I was teaching.... That was a PBL that was based on a two-week cycle. I presented them with typical higher education, tertiary education, teaching problems. And we met face to face in groups, facilitating groups, so it was more like the medical PBL in Aalborg, project based, identifying basic issues, and allocating tasks in the first meeting. And we did not meet again for two weeks, and they located the resources and posted them on the website. And then we came back together. We spent the first half of the meeting pulling it all together, finishing that problem, and we started the next one. That was much more like a medical type of PBL with a relatively short cycle."

The staff member tended to link medical PBL to the Mac Master PBL model (Barrow, 1986) where the students work on a medical case in groups. Indeed, PBL originated from the medical field and the medical PBL has so far become one of the most representative PBL models in the PBL domain.

Nevertheless, the managers at XU, though agreeing upon the notion of PBL within the medical field, emphasized that there is a distinct difference between the medical PBL and the engineering PBL:

> "Other schools don't have PBL in the same way that engineering does. The nursing field works very much on case studies, in a very highly simulated environment. So we have very high technology, a digital human being that has blood and blood pressure. So they do some of their work there, and they do other work in clinical settings. For nurses they have to do about 900 hours in a hospital- and community-based setting, as well as simulated learning activities... You might say that that is problem-based learning as well. But when you come to talk to us about problem-based learning, we are going to focus on the engineering because we think that it is a whole-curriculum approach."

From the point of view of the managers, medical PBL, though proven in the medical domain, cannot be used as an official PBL approach in engineering programs at XU, where the term PBL refers solely to an engineering PBL approach. It can be further inferred that other types of PBL should not be encouraged.

7.1.1.2 Different interpretations of PBL at XU 2: PBL as problem solving

Some staff members regarded PBL as pervasive in education. From this point of view, PBL is not special but serves as a basic feature of education. A senior staff member who had worked in industry for years argued:

> "XU was for a long time involved in delivering part of the content through PBL. Perhaps many other universities were also involved in it. Of course it is not called PBL. You cannot teach engineering without bringing in problems from the outside. I think the same applies to law, to accountancy, to marketing."

Another young staff member with an art background reached a similar but complementary conclusion:

> "When I did my degree, because I did an art degree, we did problem-based learning in art, which was that you are given a problem or a project, and then you work on it, so PBL has been taught in art for a very long time. Like drawing a chair."

They regarded PBL as being prevalent in educational settings, since learning always involves dealing with problems. Whether it is an industrial problem or an artistic one, they all share the similarity of dealing with problems. However, the kind of problem solving used generally in education cannot readily be regarded as synonymous with PBL because it involves only the elements of problem solving, that is, the application of knowledge to a specific problem setting. The acquisition of knowledge comprises the primary learning objective in this paradigm, and the application of knowledge only serves as a means to facilitate and test the students' understanding of knowledge from the lecture.

A senior staff member, though acknowledging many engineers' experiences in problem solving, refuted their attempts to transplant their industrial experience to the university setting and equate PBL with problem solving in industry:

> "Engineers, they know about problems. I think they know about problem solving. When you talk about problem-based learning or project-based learning, they think of it as the same as problem solving. So they think that if you give a series of lectures or something, then you get the students to apply it to a problem. That is problem-based learning. So we have lots of arguments, but one of the key characteristics of problem-based learning is that it starts with a problem."

From this quotation it can be seen that the key trait distinguishing PBL from industrial problem solving is that PBL sees the problem as the departure point for learning, whereas the industrial setting views the problem as a place to examine the already acquired knowledge. Sometimes these two models are treated as synonymous, but there have also been a great many attempts to differentiate between these two concepts (Savin-Baden, 2003). Different interpretations of problem-solving learning, whether it is perceived as differing from PBL or as synonymous with PBL, may cause tensions in the organization. For example, staff members who see problem-solving learning as synonymous with PBL may think that they have already been delivering PBL. Those who tend to view PBL and problem solving as two entirely different approaches may be critical of that viewpoint.

7.1.1.3 Different interpretations of PBL at XU 3: PBL as managerial strategy

Although PBL was officially introduced at XU for the first time during the time period covered in this study, it was not totally alien to all of the staff members working at the university. As noted above, some staff members had some knowledge of medical PBL, or had been involved in teaching and learning using problem-solving techniques, which they believed to be a form of PBL. Therefore, a great many members of the staff already had some opinions on the interpretation of PBL that differed from the officially implemented interpretation.

As noted above, managers noticed the existence of different forms of PBL. From their point of view, XU's PBL model was not the unique PBL approach but worked well within the engineering field. Therefore, it was an officially established PBL interpretation in a particular context. Many staff members with their own opinions of and experience with PBL perceived the official PBL interpretation as a managerial strategy to standardize their behavior. As a senior staff member commented:

> "XU was for a long time involved in delivering part of the content through PBL. . . . I guess what happened two years ago was that the university wanted to formalize this and make sure that we do it correctly."

It can be inferred from the quotation that PBL implementation at XU was not about introducing something new; rather, it was a matter of managerial intention to ensure the correctness of the organizational behavior. In this process, the old practice of PBL, which was not consistent with the official interpretation, would be corrected, replaced, or even eliminated, which could mean potential further tensions in the organization.

7.1.1.4 Different interpretations of PBL at XU 4: PBL as a whole curriculum approach (EE PBL model)

The School of Electric and Electronic Engineering (EE) PBL model views PBL as a whole curriculum approach integrating and coordinating different subjects through the means of a large project. As we noted previously in chapter 5 (XU story), the EE PBL model consists of several traditional subjects that support the project work by offering basic knowledge content and a large project where students work on a common project in groups. As part of this systematic approach, the assessment method is also adjusted in accordance with the learning objectives. The principles of PBL (Kolmos, 2008) are manifested in this PBL approach. Problem centeredness is manifested by the fact that the problem lies at the center of the curriculum design in terms of its proportion of and position in the curriculum system (if we treat the notion of a problem as relatively synonymous with the group project). Interdisciplinary learning can be seen in the project work, which aims to coordinate different disciplinary subjects. Social learning is also included, as the students are required to work on the project in groups. Moesby (2004) further discussed the characteristics and benefits of this whole-curriculum approach, which requires a holistic change affecting the entire organization, especially a change in the organizational culture. This approach differs from when PBL implementation is implemented only at the individual or the course level, in that in such a case the organizational level is barely changed.

7.1.1.5 Different interpretations of PBL at XU 5: PBL as a course-level approach (ACME PBL model)

Overall, the proportion of PBL in the curriculum at both EE and ACME is the same, comprising a 50 percent of the required units. However, the underlying principles of the two approaches are significantly different from each other. The School of Architecture, Civil and Mechanical Engineering (ACME) PBL model does not coordinate its subjects through the use of one big project. Rather, it favors the use of several small projects, each of which is linked to a particular course. Although the ACME model uses projects to coordinate its knowledge content, they are confined to the intra-course level, whereas the project in the EE PBL model serves to integrate different disciplines and thus facilitates interdisciplinary learning. The principles of PBL are partially manifested in the ACME PBL model: problem centeredness can be seen, since the project still lies at the center of the curriculum and makes up 50 percent of the curriculum system. Students are encouraged to work in groups, which can be viewed as social learning. The notion of interdisciplinary learning is not so explicit in this PBL model, since such small projects are unlikely to integrate different disciplines.

7.1.1.6 Different interpretations of PBL at XU 6: PBL as a whole-curriculum approach (XU 2010 PBL model)

Like in EE, the 2010 PBL model at the School of Engineering and Science takes a whole-curriculum approach to PBL. The new model continues to use a big project to integrate different subjects. However, one major difference from the EE PBL model is that the new school-wide PBL consolidates and strengthens the status of LBL in the curriculum system. The subjects of math and physics make up a larger portion of the curriculum—going from 25 percent in EE PBL to 50 percent in the new model. The fundamental impetus to increase the proportion of PBL is to highlight the importance of the acquisition of the basic knowledge content.

7.1.2 Review of the Interpretations of PBL at XU

The fourth, fifth and sixth interpretations of PBL at XU can be classified as managerial interpretations of PBL put into practice. It is worth noting that these managerial definitions of PBL differ in the function of the problem and the importance of basic knowledge. According to the standards of Moesby (2004) and Kolmos (2008), the EE PBL and the 2010 PBL can be regarded as better approaches because they employ a systematic approach to transforming the whole curriculum system, whereas the ACME PBL functions only at the course level, failing to change the organizational culture or manifest the principle of interdisciplinary learning (Kolmos, 2008). However, some staff members argue that the smaller-project model is well suited to the ACME context. Despite their differences, these interpretations also share some commonalities. All of these managerial approaches, whether they are based upon the design of a big project or smaller ones, can be categorized as the "*hybrid model*" (Armstrong, 1997; Savin-Baden & Major, 2004), the blending of PBL and LBL. On the one hand, these approaches encourage the use of PBL. On the other hand, they still highlight the importance of basic knowledge content by allocating a minimum of 50 percent of class time to LBL, which offers the students fundamental knowledge content and helps them consolidate their knowledge basis. Furthermore, the blended use of PBL and LBL can be viewed as a compromise between two distinct educational objectives: knowledge retention, and skill development and attitude cultivation. It can be further inferred that it is quite difficult to run a pure PBL model in practice, since students always need a foundation of basic knowledge in order to work on a problem. Given this point, mixing PBL and LBL can be seen not only as a compromise between different educational objectives but also as a compromise between two ways of knowing. Social constructivist theory views knowledge as being constructed through interaction between learners. However, the pure constructivist view does not work in practice, so there must be a compromise between social constructivism and the imparting of knowledge. The result of this compromise is the mixed use of PBL and LBL.

7.1.3 Different Interpretations of PBL at ZU

At ZU, since the university management did not impose a single PBL interpretation throughout the organization and encouraged departmental and individual experiments, many variations of PBL emerged. These can be classified as follows:

7.1.3.1 Different interpretations of PBL at ZU 1: PBL as part of the delivery approach

In some courses (e.g. Physiology, Cardiology, Gynecology etc.), part of the knowledge content is delivered through the students' work on the problem, but the rest is still delivered as LBL. Given the differences in time, resources and energy in different departments, the proportion of PBL in the courses varies from 10 percent to 100 percent.

The use of PBL as part of a course is often associated with the viewpoint that PBL cannot replace LBL. A great many staff members insist that PBL and LBL complement each other and serve different purposes: PBL is advantageous in developing students' communication skills, problem-solving skills and self-learning skills, whereas LBL's strength lies in advancing students' basic knowledge and consolidating their knowledge foundation.

It is worth noting that PBL and LBL contribute to different educational objectives. It is necessary to keep LBL in the educational system as long as knowledge acquisition is still considered to be one of the educational objectives. As many studies suggest, LBL is especially powerful in improving students' short-term memories (Strobel and van Barneveld, 2009). For these reasons, LBL cannot be eliminated from the current educational system.

7.1.3.2 Different interpretations of PBL at ZU 2: PBL as a whole course approach

This could be seen as a pure PBL approach, though it only exists at the course level. In these pure PBL courses (e.g., Pathophysiology), lectures are totally eliminated from the course, and all the knowledge is gained through the students' work on the medical problems. This approach is employed only in some clinical courses.

7.1.3.3 Different interpretations of PBL at ZU 3: PBL as a cross-course approach

Cross-course PBL occurs in integrated courses at ZU (e.g., Bio Chemistry). Theoretically, the cross-course PBL approach is aimed at building a link between different subjects through the process of dealing with a big case. In this sense, this approach can be adopted to facilitate interdisciplinary learning (Kolmos, 2008), as students should be able to recognize the connections among various disciplines and work on the problem drawing on knowledge and methods across disciplines. As we noted previously, XU's whole-curriculum approach incorporated interdisciplinary learning by coordinating various subjects in the project work. However, there are a limited number of interdisciplinary learning elements in the course-level PBL approach at ZU, because although integrated courses combine several related subjects, the boundaries between these subjects can still be identified. To put it another way, the subjects in these courses are hardly integrated. Rather, the subjects are taught in parallel, with distinct teachers, course content arrangements and assessments. This system hampers interdisciplinary learning.

7.1.3.4 Different interpretations of PBL at ZU 4: PBL as project-organized learning

In some departments (e.g., Bio-medical Engineering), PBL is interpreted as project work for which students are required to work on a project or task in groups. However, the project work is designed based on a single discipline, and the number of interdisciplinary learning elements in the project work is quite limited. The project work at ZU is similar to that at XU, but only runs at the course level.

7.1.3.5 Different interpretations of PBL at ZU 5: PBL as discussing a social or disciplinary issue

In the social science departments (e.g., Medical Ethics), PBL is often interpreted in such a manner that the students discuss a social or disciplinary issue in groups. Unlike working on a project or dealing with medical cases, the students using this PBL approach are normally not required to finish a task (in medical PBL, students usually have to complete a group presentation) or make a product (as in project work). This approach highlights the process of learning rather than the product or final result. The students are expected to deepen their understanding of the social and disciplinary issue through challenging and reflecting on each other's viewpoints.

7.1.3.6 Different interpretations of PBL at ZU 6: PBL as question answering

The simplest way to understand and conduct PBL that we have noted so far is to equate PBL with addressing the questions posed by the teacher. Due to time and resource limitations, many staff members argue, they cannot afford to design a medical case or project. Therefore, they have to use the most cost-effective way of conducting PBL, which has led to the emergence of this approach. PBL questions are open-ended and are posed for the purpose of exploration rather than to solicit a particular answer.

7.1.3.7 Different interpretations of PBL at ZU 7: PBL as a lecture-based case

This approach is similar to the one described by Barrow (1986) in that a case is used as a medium to explain the information in the lecture. The information is given first, and then the lecturer presents the case to illustrate the information. The students stay passive throughout the whole learning process. As one teacher commented, "*PBL is a new way of teaching; it can help me to use a case to explain the learning material to the students, who are then able to see how the knowledge is applied in a particular context.*" The case study is used as a teaching technique to assist students to better understand the learning material. The learning process does not change much, since the teacher still dominates the teaching and learning process.

7.1.4 PBL Interpretations at Various Levels

Based on the preceding analysis, it can be seen that there are a number of PBL interpretations at both universities. The variations of PBL can be fit roughly into three different levels:

1) Managerial (university level) interpretations are referred to as the officially defined PBL and often determine how PBL will be implemented throughout the entire organization. Furthermore, the managerial interpretation of PBL is readily found in the literature, as it is the valid definition at the university. At XU, the managerial interpretation of PBL comprises the whole-curriculum approach, in which half of the curriculum is dedicated to traditional subjects while the other half is made up of project work that brings together different subjects. At ZU, the detailed organizational interpretation of PBL is absent because the senior managers have decided to encourage the development of multiple PBL definitions at the bottom level.
2) Departmental interpretations of PBL refer to PBL interpretation in a particular department, faculty or other academic unit of the university. The departmental interpretation of PBL is only valid for a particular academic unit and is not implemented in other academic units at the same university. Therefore, several departmental interpretations of PBL are likely to exist simultaneously at one university. Prior to the introduction of

the 2010 PBL model, there were two departmental PBL interpretations at XU, namely, EE PBL and ACME PBL. At ZU, since there is no university-level interpretation of PBL, most existing PBL forms can be perceived as departmental PBL interpretations.

3) Individual interpretations of PBL refer to how individuals (including both managers and teachers) understand PBL in practice. Although there is little systematic record of individual interpretations of PBL, the individual understanding of PBL is crucial for PBL implementation, since it is the individual who translates PBL into real practice and has an actual impact on students.

These three levels of PBL interpretation are not always consistent with each other; rather, they differ from each other in their prioritization of educational objectives, organization of PBL curriculum and allocation of limited resources for PBL.

7.1.4.1 Pure PBL and hybrid PBL

Pure PBL sees knowledge as being constructed through students' self-inquiry and their interaction with other human beings. It also emphasizes the importance of skill development and attitude cultivation in its educational objectives. Hybrid PBL, the mixed use of PBL and LBL, on the other hand, admits the value of LBL and knowledge acquisition to a certain extent. The emphasis on the importance of basic knowledge acquisition grows as the proportion of lecture in the curriculum system increases. It is worth noting that the majority of the interpretations of PBL, both managerial and departmental, belong under the umbrella term hybrid PBL.

7.1.4.2 PBL as a whole-curriculum approach or as a course-level approach

When PBL is seen as a whole-curriculum approach, it is often adopted in the form of a big project coordinating different subjects, as in the EE PBL model and the 2010 PBL model at XU. In contrast, the course-level PBL approach often uses small projects, each one based on a single discipline. The cross-course PBL at ZU seems to lie between the whole-curriculum PBL approach and course-level PBL. In general, whole-curriculum PBL seems superior because it manifests the principles of PBL better than the course-level PBL does. For instance, the whole-curriculum PBL approach involves more elements of interdisciplinary learning than the course-level PBL approach because the big project is able to integrate various subjects.

7.1.4.3 PBL interpretation that deviates from what is described in the literature

The aforementioned notions of PBL (hybrid PBL, pure PBL, PBL as a whole-curriculum approach or as a course-level approach) are widely recorded in the literature (Savin-Baden & Major, 2004; Moesby, 2004). In practice, we notice that various interpretations of PBL, most of which are individual definitions, are inconsistent with what is widely recognized in the principal theories. For instance, the tendency to equate PBL with problem-solving learning appears to contradict the argument that they are two entirely different approaches to learning (Savin-Baden & Major, 2004). The understanding of PBL as groups discussing social or disciplinary issues, or as addressing the question posed by the teacher, is inconsistent with PBL principles, as noted previously. Nevertheless, we should not underestimate the influence of the individual interpretation of PBL, since it is the individual teacher that controls the actual impact on students through daily teaching practice. Furthermore, we should caution against any attempt to treat these definitions as something that should be eliminated from the educational system due to their inconsistency with the essence of PBL. The reason for caution lies in the fact that sometimes these definitions may be the best practical compromise under conditions such as a limited amount of time and resources.

7.1.4.4 Knowledge and beyond

The diversity of PBL interpretations can be accounted for by the multiple viewpoints of the value of knowledge retention in education. For example, hybrid PBL assumes that PBL can only comprise part of the curriculum because a considerable proportion of the education system should be based on LBL. Skill development and attitude cultivation cannot be facilitated without the necessary accumulation of basic knowledge. On the other hand, pure PBL is based on the assumption that all necessary knowledge can be gained through students' self-inquiry and interaction among students. Based upon our interviews, it is clear that knowledge acquisition is still emphasized by a great many staff members and students. Knowledge acquisition is important for the following reasons: first, it serves as the basis for higher-level cognitive activities. The hybrid PBL model defines the lecture as a support to project work. Second, the assessment method still focuses on the retention of knowledge, which partly dominates the students' learning process. Finally, it is believed that a solid knowledge base will be conducive to future professional development.

7.1.5 Reasons for the Diversity of Interpretations of PBL

As to why there are so many different interpretations of PBL at a single university, we refer to Gadamer's (1995) notion of "fore-structure" in order to grasp how human understanding takes place. The "fore-structure" refers to a person's cognitive structure before he embarks on the understanding activities. It becomes the condition for the understanding, making it possible and, at the same time, largely confining it. This section does not attempt to make a thorough study of the notion of "fore-structure"; rather, we are more interested in the relationship between the interviewee's fore-structure and his interpretation of PBL. The implication is that the study should concern itself with the staff's educational experience and working experience, because it is these elements that influence a person's fore-structure, which further creates his/her perception of PBL.

It is clear that an individual's perception and interpretation of PBL can be linked to the interpreter's own experience. Staff members tend to view PBL as synonymous with problem solving due to their work experience in industry; a teacher thinks that PBL is prevailing in the educational domain because he experienced a lot of problem-solving situations in his educational experience; a student who highlights the importance of knowledge acquisition at university may have gotten used to traditional educational concepts during his primary and secondary school experience. At XU, a manager attributed the reluctance of staff to conduct PBL activities to the university's history:

> "This institution was a teaching institute, so there was very little research, it was just teaching, in pretty much the traditional way of teaching. So the staff members in our faculty are here from that type of background, they are not researchers. If you are a researcher, if you have got experience in research... you are not frightened of leading a project where you have to answer all sorts of crazy questions... because you are not showing them what the answer is and teaching it point by point.... [Old staff] don't have wide experience in research... they feel a little insecure because they are afraid of students, they are frightened of discovering because they might be not able to answer the question."

Since staff members as well as students vary in their educational background, work experience, or, more common still, their personal life experience, the diversity of PBL interpretation cannot be avoided. Each of them will bring their own fore-structure into the PBL-understanding process and thus create variations of PBL.

7.1.6 Tensions and Chaos

The diversity of PBL interpretations may cause tension when the university employs a normative approach to impose the organization's interpretation of PBL on teachers and students. At XU, the inconsistency between the different PBL interpretations can be readily recognized. As we saw in the XU story, there was an appreciable level of resistance to the use of PBL, which was widely considered to be a huge challenge for the change process. This resistance can be interpreted in various ways: it can be regarded as a tension between LBL and PBL, since there were many staff members who were not willing to accept PBL and stuck to LBL; or it can be understood to be a conflict between different interpretations of PBL. For instance, the EE staff and ACME staff at XU held quite different conceptions of what PBL is. In addition, it can be also considered as tension between the management's efforts and the perspective of staff members, who had their own opinions of PBL. For example, some staff members viewed PBL implementation at XU as a managerial attempt to standardize the staff's behavior, as we saw in the third interpretation of PBL at XU. As long as the organization wishes to impose a single official PBL interpretation, the tension among various PBL interpretations cannot be fully eliminated.

At ZU, since the management did not impose a single university-level PBL model, various forms of PBL interpretations emerged and developed. There did not seem to be much tension among different PBL interpretations, since there was no PBL interpretation imposed from above to replace others; however, the situation became a little bit chaotic. Some of the PBL definitions are far away from the principles of PBL, such as the fifth, sixth and seventh PBL interpretations at ZU. In addition, since there was a lack of systematic design, different departments and individuals had the freedom to decide on the schedule of PBL activities, which led to further chaos. First, the lack of managerial attempts to coordinate different departments and individuals meant that, at a given moment, there might be several PBL courses occurring at the same time, which created a huge academic burden for students, and during some periods there might be no PBL activities at all. Given this problem, some staff members suggested that there should be some managerial attempts to coordinate PBL activities of different departments and individuals so as to provide the students with a more coherent, more orderly and more effective PBL experience. Furthermore, although there were some high quality PBL interpretations, such as those who saw PBL as a whole-curriculum approach or a cross-course approach, we still noted that some PBL interpretations, such as the one equating PBL to a lecture-based case study, were a far cry from the principles of PBL and would need to be modified in the future. Although the staff might claim that they were conducting PBL, they probably implemented it in completely different ways. Therefore, it is necessary for those at the management level to take measures to secure the quality of PBL by examining and reflecting on those PBL approaches that are not of such high quality.

7.2 Interpretations of Social Learning

The preceding section discussed the organizational members' understanding of PBL in general. It is worth noting that the majority of the above discussions focus on the design of the curriculum and the learning process. In this section, we turn to another essential aspect of PBL, social learning, which also shifts our attention from the organizational level to the micro level, that is, how PBL actually functions in the classroom.

There are many attempts to address how to design the learning process at the micro level, how students experience PBL and which learning strategy students tend to use. However, there is not much research regarding how PBL really takes place at the micro level, and it is for this reason that we now focus on the extent of the interaction among group members.

Social learning is widely recognized as one of the major traits of PBL (Kolmos, 2008). In general, social learning is defined as a learning approach where students work on the problem situation in a group. It is believed that through interaction, mutual challenging, presenting and communication the students are able to develop their communicative skills and teamwork skills. For the purposes of this study, as we pointed out in the theoretical and methodological section, this study includes the tutor as an integral part of social learning. Unlike the interpretation of PBL, which can generally be recognized at both the systematic level (curriculum design) and the personal level (individual interpretation), social learning mainly happens at the micro level, for instance, the classroom. Our data for social learning is mostly drawn from the interviewees' reflections on their own learning experiences as well as our observations of classrooms and other learning environments.

7.2.1 Interpretations of Social Learning

In general, based on our findings, social learning happens in the following forms:

7.2.1.1 Interpretation of social learning 1: In the group (without clear role distribution)

A group is assigned a problem to work on. Sometimes, students are required to conduct some individual inquiries beforehand and then bring their findings and thoughts to the group discussion. Since the majority of students are relatively well prepared, there are always interesting dynamics in the group discussion, such as we observed in the class on medical ethics. Students take turns to present their own thoughts and question and challenge others' viewpoints.

7.2.1.2 Interpretation of social learning 2: In the group (with clear role distribution)

Group work can take other forms, as well. At times, the group used task distribution among its members. Students in the group took on different roles, such as the host, recorder, observer, etc. Although the role distribution was necessary, in some groups, the role distribution among the group members hampered the interaction of the students. Sometimes, as the interaction among the students took a lot of time and energy, since the students needed to prepare for the discussion, the students tended to use the role distribution as a way to reduce or avoid the large amount of effort they needed to invest in the group work. In this type of group, where the students' main focus is on their own role or task, the interaction among students is quite limited.

The extreme case of task distribution happens when the group adopts a "take turns" strategy where group work is assigned to a minimal number of students in the same group. For instance, if a group of 10 students has five problems to work on in a semester, the group distributes five problems among the 10 students in same group in such a way that every two students are responsible for one problem. In this case, interaction and communication among the students in the group is reduced to a minimal level: each student is only concerned with his or her assigned problem and thus is totally ignorant of what the other problems are and what other group members are doing.

7.2.1.3 Interpretation of social learning 3: Among the groups (with clear task distribution)

Big-class PBL usually uses a strategy of distributing tasks among different groups. Each group is assigned a problem to work on and they present their findings to the large class during the large class discussion procedure, as occurred in the cell biology class. Each group is only responsible for its own problem and is therefore ignorant of what other groups are doing. Due to the large size of the class, the large class discussion is similar to a traditional lecture, where the interaction among the students is quite low. The only difference between this class and LBL is that it is the student, rather than the teacher, that is responsible for delivering the knowledge content to other students.

7.2.2 Tutor in Social Learning

In most of the PBL classes we observed the tutor was minimally participative in the learning process. To put it in another way, the tutor had the tendency to leave the students alone during group work even though he or she was present. The minimal participation of the tutor in students' learning is not consistent with the essence of social learning, as the tutor is required to be actively involved. The teacher in the PBL environment seems to go to the other extreme with regard to the student-teacher relationship. Students should not be treated as passive learners with little freedom, as in LBL, nor can they be treated as learners without supervision. Rather, the student-teacher relationship should be treated in a dialogical manner where students, though they may be directing the learning process, require equal participation from the teacher.

Attention is always an issue for large-sized groups where the tutor cannot give enough attention to all the students or groups (Shipman & Duch, 2001). It is indeed a big challenge for PBL. However, this section is more concerned with the problem of role/task distribution in PBL groups, which is seldom discussed. Role distribution is regarded as a significant strength of PBL groups, since it enables students to collaborate in a more effective way. However, there are also some drawbacks to it.

7.2.3 Group Effectiveness and Learning Process

Role distribution in groups reflects the impetus to promote group effectiveness. Though PBL highlights the importance of the learning process rather than the learning product, learning always involves the production of some outcomes.

The outcome could be a group report, a group presentation or even a product. As long as the learning involves producing several outcomes, the students are likely to take measures to enhance their work effectiveness in the group so as to accomplish the given task. Thus the learning objectives are displaced by the motivation to finish the task. Since a great many students view communication and interaction among group members as taking up a large amount of time and energy, they are likely to distribute roles among group members and therefore reduce the elements of social interaction

7.2.4 Classroom Effectiveness and Learning Process

Making role distribution among various groups is based on the staff's intention to enhance classroom effectiveness. As a staff member commented:

> "We have a great deal of knowledge content to deliver to the students. We have to cover all this content in our class. It is impossible to let every student experience all the problems, so we have to assign the problems to different groups. Otherwise, we are unable to finish our teaching and learning goals."

Another staff member put it in a similar manner:

> "It is ideal to offer the opportunity to every student to engage in the discussion and presentation; however, we have too many students, and we cannot afford that. We have to make sure that we finish the teaching content first."

Staff members struggle to balance classroom effectiveness and the learning process, which seem to be in conflict with each other. Classroom effectiveness is largely concerned with delivering the intended amount of knowledge to the students, from the point of view of staff members. PBL is considered to be an obstacle to classroom effectiveness and detrimental to the accomplishment of educational objectives. Therefore, the staff has to employ the role distribution strategy to ensure that the teaching objective, or the prescribed amount of knowledge, can be achieved in time.

PBL places significant emphasis on the learning process rather than on the product. The social learning theory also supports a similar emphasis and thus encourages concerted efforts among the students. However, intensive communication among the students would require a large amount of time and energy, which neither the students nor the staff can afford. Moreover, both students and staff are under pressure to complete the prescribed task on time. For the students, the outcome could be a group report, a presentation or a product. Staff members have to make sure that the prescribed body of knowledge has been delivered to the students. Therefore, both the students and the staff have to ensure the effectiveness of both the group and the class, which is intrinsically contradictory to PBL's emphasis on the learning process. The tension between PBL's focus on the learning process and the organizational members' (teachers and students) intention to guarantee educational effectiveness largely constrains PBL implementation because PBL is aimed at the learning process rather than the completion of learning tasks.

7.2.5 Student-authority Relationship

LBL is characterized by the fact that students passively receive the information delivered by the authority. The authority can take various forms, for instance, a teacher, a textbook or even another student. When discussing the issue of dialogical learning, or the restructuring of the student-authority relationship, we refer not only to the transformation of the relationship between the student and the teacher, but also to the shift of the relationship between the student and other forms of knowledge authorities such as the textbook. In this sense, the dialogical approach suggests that textbook knowledge should also be seen as an object to engage with in dialogue. In our cases we note that although the teacher, the traditional knowledge authority, may be absent from the class, other knowledge authorities are still present. For example, one of the typical ways of conducting large-class PBL is to assign a problem to a group that must then present their findings and thoughts regarding the problem to the rest of the class. Due to the lack of interaction in a large class and other groups' ignorance of the problems that are not assigned to them, the classroom becomes a mini lecture where a group of students make presentations to the class. The majority of the students in a class tend to conform to the authority represented by the presenter who is discussing his/her group's findings regarding a particular medical problem. The relationship between the student and the knowledge authority does not change at all. PBL in these circumstances only leads to the altering of the knowledge authority from the teacher to other students (who make presentations), rather than the restructuring of the student-authority relationship.

7.3 Review of the Understanding of PBL and Social Learning

Since the last century, there has been a great paradigm shift within the educational field, from an *instruction paradigm* to a *learning paradigm* (Barr & Tagg, 1995), implying that educators in higher education institutions should shift their attention from teaching to learning. In this sense, a university with a learning paradigm is aiming at producing learning, in contrast with the instruction paradigm, in which the university is a place offering quality teaching. Within this context, the university should endeavor to provide a learning atmosphere rather than simply conveying the knowledge to learners. PBL can be regarded as an effort to facilitate this educational paradigm shift. By paradigm shift we mean that PBL refers to a radical change of our conception of knowledge and learning. PBL is more concerned with learning rather than with teaching or curriculum design; PBL views knowledge as being constructed by the learner rather than transmitted from external authorities; PBL sets problems at the center of learning; PBL emphasizes the thorough exploration of a single case through practice, rather than the learning of general principles; PBL focuses on the learning process rather than the product; PBL represents a new way of learning and knowledge and should not be simply treated as a technique; PBL refers to a dialogical relation between students and knowledge authorities. In our cases, it can be seen that throughout the change process to PBL, PBL has become an organizational discourse; however, whether the paradigm has shifted remains questionable.

Firstly, all the interpretations of PBL focus on the design of the curriculum and teaching rather than the understanding of the students' learning process. Whether it is a managerial interpretation of PBL, departmental interpretation of PBL, individual interpretation of PBL, hybrid PBL, pure PBL, problem solving, case analysis, etc., it represents the viewpoints of the teachers and the curriculum developer at the university rather than those of students. The managerial (institutional) interpretations of PBL are more concerned with curriculum design and therefore view PBL as different types of PBL curriculum models, as we noted earlier. As for the staff, they are more intrigued by the design of the instructional process, such as how to design a good problem or a medical case; they are, to some extent, lacking in the understanding of learning. If we come back to the term PBL, we see that PBL is composed of both P (problem) and L (learning). The 'P' of PBL is the means to facilitate learning; the 'L' of PBL is where our focus should be. It seems that both managers and staff members pay too much attention to the 'P' component, and thus they more or less neglect 'L' (learning). In this sense, we can argue that at both universities, the organizational members are, to some extent, dealing with PBL using an approach belonging to the *Instruction Paradigm* rather than the *Learning Paradigm*. That is, they are more concerned with teaching than with learning. As noted above, PBL should shift the educational focus from teaching to learning, since PBL places students at the center of the whole learning process. Here we are not asserting that one should abandon teaching and curriculum design when implementing PBL, because PBL still requires a design dimension for the curriculum and instruction process; rather, we should transcend the discourse of the instruction paradigm and pay increasing attention to the learning of the students when dealing with PBL.

Furthermore, staff members at ZU maintained that lecture is an indispensable part of the curriculum system. They contended that the lecture is "*more effective than PBL*" in delivering knowledge to the students. Some students shared similar arguments that PBL is not very efficient in helping students acquire the basic knowledge in the course. In addition, some students tend to have difficulties in understanding some learning materials when learning on their own. Therefore, they resort to traditional lectures in which the teacher offers them clear answers. In this sense, both staff and students insist that although PBL is rewarding, it cannot replace traditional lectures. Furthermore, at both universities, we see a great many staff members and students who stick to the traditional educational approaches and thus do not recognize the value of PBL. Therefore, dialogue between PBL proponents and opponents is necessary and will take longer than expected.

Moreover, although the ideal of PBL is to emphasize the learning process rather than the learning outcome, there are always some ends that PBL has to accomplish. These learning goals could be an individual report, a group presentation, a mini product, etc. Due to the curriculum schedule, students have to finish the required tasks in a given time. Given the limitations on students' time and energy, they are likely to develop their own time-saving strategies in order to complete the assigned tasks. From my previous description of the group work at ZU we can see that a considerable number of students adopted the role distribution strategy in their PBL class. Although it is highly effective in finishing a given task, role distribution largely inhibits students' interaction and communication, which is widely considered a basic principle of PBL. Under this strategy, the emphasis on learning processes is intentionally replaced by emphasis on learning products. Since the learning outcomes at ZU are mainly the acquisition of knowledge, it can be concluded that the focus on knowledge retention surpasses the concern for the learning process.

Many staff members and students recognized the merits of PBL, which are often given as the justification for the use of PBL. It should be acknowledged that the merits of PBL have been widely recognized. For example, PBL is beneficial in promoting the learner's knowledge acquisition, generic skills, self-inquiry learning skills, self-directed learning skills, life-long learning attitudes, and so on. The danger of identifying these areas of growth is that organizational members may perceive PBL as a technique to achieve some educational objectives. When it serves only as a technique, PBL has to subject itself to educational objectives. However, PBL is far more than an educational technique that produces some educational benefits. PBL should not be conceived of as simply a technique; rather, we would argue for PBL to be used in an ontological sense. Or, in one researcher's words, PBL should be seen as a philosophy guiding the whole educational process (Margetson, 1991). In this respect, PBL refers to an alternative way of viewing knowledge and learning. Knowledge is not conveyed from the external environment; rather, it is formulated through the learner's self-inquiry and construction. Learning is not an individual process; rather, it is a social process involving interaction among many learners. PBL involves dealing with particular case-like problems, and it is through dealing with particular cases that the learner is able to transcend rule-governed rationality and gain genuine expertise (Flyvbjerg, 2006). It should be noted that although the ontological meaning of PBL for education has been recognized in a great many theories, the staff members in higher educational institutions are not fully aware of this. Their praise of PBL is based upon its instrumental values, such as "PBL can provide us something that the traditional approach cannot offer," or "PBL can help our students to develop their soft skills," and so on. This partially explains why the university does not change its assessment method, since staff members view PBL as a new teaching method to improve students' learning outcomes.

The attempt to reduce PBL to its instrumental strength largely underestimates its value. As we noted earlier, PBL represents a new way of viewing knowledge, learning and knowing. In my interviews, we see that the staff members at XU and ZU, despite being aware of the benefits of PBL, seldom mentioned PBL from the perspective of conceptual change. Furthermore, the tendency to reduce PBL to an instrumental level may be destructive to the staff's ability to make a conceptual change. Recognizing the merits of PBL does not necessarily mean that staff members and students change their conception of learning. Their conception of learning is even more unlikely to change when the assessment method still sets traditional educational goals as its focal point. Furthermore, the relationship between the staff's conception of teaching and learning and their teaching practice is quite unclear. The knowledge of PBL does not necessarily include the dimension of action. The conceptions of both learning and teaching held by teachers have an impact on their approaches to teaching (Trigwell & Prosser, 1996; Trigwell, Prosser & Waterhouse, 1999). However, there is a notable gap between the teacher's view and actions, implying that a constructive view of teaching and learning held by a teacher does not guarantee a constructivist-oriented teaching style (Fischler, 1994). Students are unlikely to change their learning approach without a conceptual change in education. It is revealed that there is a relationship between students' perception of learning and their learning approach. Students who see learning as increasing knowledge or retention, that is, who see learning as external to themselves, tend to use surface learning approaches. They are quite concerned with satisfying the requirements of the assessments. On the other hand, students who see learning as an internal process, or meaning construction, tend to use deep learning approaches (Ramsden, 1992; Turner, 1995).

Within the PBL environment, problem centeredness should be adopted as a basic principle for the design of the curriculum; that is, the curriculum should be centered on problems. Thus curriculum design should start from the consideration of problems rather than from concrete disciplinary knowledge. At ZU, however, we see that the curriculum is designed on the basis of disciplinary knowledge. Afterwards, part of the curriculum is made into PBL to facilitate students' understanding of the information in the lectures. Attempts to design a curriculum based on disciplinary knowledge and then make it problem-based usually lead to disaster (Savin-Baden & Major, 2004: 45). When the curriculum is designed on the basis of traditional subjects and then made problem-based, the problem is more likely to serve to apply the knowledge that the students obtain from the lectures to the practice.

It is worth noting that prior to this study it was supposed that any obstacles to PBL implementation were likely to come from staff members. At a university, students are just temporary organizational members, whereas the staff members are the permanent ones, and thus the students play a relatively inferior role in the change process compared to the staff. Nevertheless, based on our empirical data, students can serve as a significant obstacle to PBL implementation. Teachers sometimes lecture not because they wish to do so, but because students require it of them. Some students insist that lectures are necessary because sometimes they are unable to "grasp the major knowledge points" from group work, and therefore they need the teacher to make a presentation so as to clarify what knowledge content they need to memorize.

Some students from PBL classes even borrow the notes from their friends in non-PBL classes as a reference, since they suspect that they have not retained as much knowledge as their non-PBL peers have. In this sense, the students at ZU do not always favor the use of PBL.

In light of the preceding discussion, we can see that although both universities have transformed their curricula to PBL, the organizational members' understanding of knowledge and learning have not fully changed. Therefore, it remains a challenging task to change the organizational members' (including managers, staff and students) perception of knowledge and learning in accordance with the concept of PBL.

8

Motivation for the Change to PBL

The motivation of the organizational members plays a vital role in any type of organizational change. There are many theories regarding human motivation, as we noted earlier. In this chapter, however, we are not attempting to explore in detail how the organizational members should be motivated by the organization to participate beforehand. Rather, we are more concerned with how the organization motivates its members during the actual process of implementing PBL, how organizational members perceive and understand the motivating issues in the organization and how they react to managerial attempts to motivate them.

The dominant theories of staff motivation focus on the role of the managers and other key actors supportive of the change. Responsibility for vision building and the formulating of a change coalition (Moesby, 2004; Cohen, 2005) seems to fall on the shoulders of managers and PBL champions. However, PBL implementation is not only the business of managers and PBL advocates, but involves the whole organization. We therefore argue that paying attention only to the management level and advocators when considering motivational issues provides too narrow a perspective. We should also be concerned with varying understandings of PBL implementation within the organization and with those members that are against PBL. In this section, the focus is not on identifying or developing strategies to motivate the organizational members to participate in PBL. Rather, we attempt to show how the university motivates its members when implementing PBL and what sources distract or demotivate organizational members from PBL activities.

8.1 Sources for Motivating Staff Members

We can see from the stories that both universities endeavor to motivate staff members to participate in the change process. From the points of view of the managers as well as some staff members, there are two main types of methods for motivating staff: economic and cognitive. The economic means are staff members' "natural impetus"—salary and promotion opportunity—and the cognitive method is primarily concerned with staff development programs.

Management at both universities provided intensive institutional support for the change. XU created a great many staff development programs, both internally and inter-organizationally, aimed at delivering PBL-relevant knowledge and skills to staff members, transforming their values and concepts and cultivating future PBL proponents. It seems that these methods gained increasing recognition from many staff members, especially those with no PBL experience. At ZU, PBL was originally introduced as part of the senior manager's economic stimulus. As previously noted, PBL was initially introduced by the establishment of a funding project for PBL. In addition, staff development programs serve as another source to provide the impetus for PBL implementation.

In addition to these methods, the existence of the power-coercive element for the change is always present. At both universities, PBL was introduced into the university by the senior management, although it was done in different manners. At XU, staff members seemed to be coerced into implementing PBL, regardless of their levels of willingness. The coercive force is felt especially keenly by those who are unwilling to participate in PBL activities. At ZU, on the other hand, although the vice president did not impose PBL, the staff members seemed willing to conform to the manager's intentions. As one staff member at ZU stated,

> "When the campaign is initiated by the president, things always go easily, even when there is no enforcement".

8.2 The Organizational System

The organizational system is sometimes inconsistent or even conflicts with the implementation of PBL. The organizational system refers to the organizational techniques used by managers such as payment, promotion and so on. The organizational system is not purely neutral and technical; rather, it includes a normative dimension that directs human behavior in a certain manner. As noted by Henriksen et al. (2004), the organization runs under a particular logic and serves particular values. It is suggested that when an organization initiates an institution-wide change it should transform its organizational system in accordance with the intended objectives. As many researchers argue (Biggs and Tang, 2007; Kolmos and De Graaff, 2007), the principle of aligning different elements in the organization should be taken into account when making a transformation. However, this is not always possible in reality. Although a university may endeavor to adjust its organizational system in favor of PBL implementation (for instance, ZU established the funding project as well as offering staff development programs), on the whole, the existing organizational system does not seem quite supportive for PBL implementation. Consider the following points that work against successful PBL implementation:

8.2.1 Research-oriented Evaluation System

As many staff members attest, research has become the dominant if not the only factor in the scaling of staff salaries and in promotion policy. Although teaching is an integral part of the staff evaluation policy, on the whole its importance has been reduced. Thus teaching evaluation has difficulties in motivating staff members to participate in teaching activities, let alone in PBL. For example, a teaching paper is seldom considered a research paper (scientific paper) in the crediting system for the staff at ZU. Although ZU has recently changed its publication policy, stating that the teaching paper could be considered a research publication, the teaching paper can only be credited as one research paper, regardless of how many teaching papers have been published. Therefore, under this research-driven policy, most staff members are directed to invest an increasing amount of time and energy in conducting research rather than in being involved in teaching activities. As some staff members commented (especially the staff members who are responsible for the teaching affairs), a great many teachers only maintain minimal involvement in teaching activities to satisfy the basic teaching requirements of the university, and thus are not very enthusiastic about teaching issues. Another manager in charge of superintending teaching affairs recalled,

> "I cannot tell you how many times I see that for a four-hour course, the teacher only gave one-hour presentations, and then left and left the students to learn on their own. The number of this kind of teachers, is, er, quite huge."

Under this general atmosphere, some PBL proponents were a little bit frustrated, as a young teacher noted,

> "Most of the teachers are not quite enthusiastic about teaching and learning. I am kind of alone."

Although some managers are attempting to reverse the research-oriented trend so as to guarantee the teacher's involvement in teaching activities, it seems that their influence on staff members is quite limited. A manager for teaching and learning affairs recalled,

> "When I train the young staff members, I tell them that the teaching is their primary mission. Research and other work are just adornments. We use the system of one-vote negation, which means that if the teacher makes a big error in the teaching, there is no opportunity for him to get promoted or get other awards. What else do you think we can do to manage the teachers? This is what we can do now. You know that in our research university, research is highly emphasized as the priority. The teaching is not given enough attention."

A senior manager maintained,

> "This is a mistake for the university. Our strength is teaching rather than research, in which we performed quite poorly compared to other research universities. Therefore, the university should locate itself as a teaching university and pay more attention to the improvement of teaching."

The lack of enthusiasm in participating in teaching is also evidenced by the fact that some teachers viewed PBL implementation as a temporary task rather than a process of changing their perspective on knowing and learning. Therefore, they did not continue to implement PBL after they had finished the program.

8.2.2 Evaluation System for Hospital Staff

A clinical orientation characterizes the evaluation system in the affiliated hospitals at ZU. Those who work at ZU's affiliated hospitals serve two different roles. On the one hand, they are teachers who give students lectures and instructions at the university, while on the other hand they are staff members at the hospital who are required to accomplish clinical tasks. These roles are not complementary, but rather they are contradictory to each other, since staff members have a limited amount of time and energy. Since the hospital has imposed a large number of clinical tasks on staff members for economic motives, the staff members at the hospital have to invest a large amount of time and energy in performing clinical tasks, and thus their time for teaching is quite limited. In addition to organizational policy, staff members' low motivation for participating in teaching activities at the hospital is also associated with their personal pursuit of financial benefit. Their compensation for clinical work is one hundred times higher than what they receive for teaching.

8.2.3 Motivation of Students

The majority of courses that include PBL as an integral part have not changed their assessment methods. The assessment methods still focus on students' retention of the basic knowledge content of the course. The assessment approach consists of blank-filling and multiple choices questions, as well as "ask and answer," which is aimed at testing students' ability to memorize information. As the principle of alignment (Biggs and Tang, 2007) suggests, PBL, or any other type of curriculum innovation, cannot be readily implemented without changing the method of assessment. Students' performance is recorded in the academic record, which serves as a crucial reference when the students apply for higher degree programs in the future or when they apply for jobs. To put it in another way, students' performance on the assessment at university may be decisive for a student's future life chances in terms of education and job seeking. Students are thus forced to try to excel in ways measured by the assessment method; that is, they have to endeavor to adopt the traditional ways of learning and to memorize the knowledge content in order to better reproduce the knowledge required by the assessment. Because of this, students are not entirely motivated to participate in PBL activities, even when they acknowledge PBL's strength in cultivating the skill set desired by employers, which includes communication skills and problem-solving skills. In our interviews we could sense the students' interest in acquiring basic knowledge. Nearly all of the student interviewees showed their conviction of the importance of knowledge acquisition in education.

From the perspective of staff members the traditional way of assessing students is more advantageous than PBL assessment, for two reasons. First, traditional assessment is rigid in its approach to evaluating the students' acquisition of course content; in contrast, PBL assessment, as many staff members commented, cannot evaluate the quality of students' knowledge retention. Again, this argument concerns assessment effectiveness, or in other words, how to efficiently judge the quality of students' learning. Furthermore, quality of learning here is conceptually associated with emphasis on the knowledge itself. Assessment effectiveness can be roughly regarded as the efficiency of the assessment method in "measuring" the quality of knowledge retention. Second, the proponents of traditional assessment argue that it is a fair method to measure students' learning outcomes without any subjective bias. If the assessment contains large amounts of open-ended elements such as oral tests, "it becomes no longer objective," as some staff members stated. From their point of view, the open-ended assessment may, in the short term, create unfairness for the students in a given semester, and it could even create unfairness in the students' future development (i.e., in seeking a job). Due to these concerns, a great many staff members are quite reluctant to change their assessment method for the students.

Students' learning approaches are influenced by various factors, such as their perception of learning, the assessment method used in the course and so on. Among these factors, the assessment method serves as a directive force for students in deciding which method they will choose in their learning process (Ramsden, 1992; Turner, 1995). When the assessment method focuses on the retention of knowledge, students are more likely to use the surface learning approach, which is aimed at memorizing content without necessarily understanding the meaning of the material learned. As noted at ZU, since the university did not make a radical change to its assessment method, many students tended to favor the traditional learning approach because the assessment continued to focus on the retention of knowledge. As one student stated:

> "Another concern is that PBL training is closely linked to practical life, but the assessment returns to the basic knowledge. We worry that we know a lot about the medical case, the clinical symptoms, but we know little about the basic knowledge."

8.2.4 National Curriculum and Requirements

At ZU and other universities in China, up to one third of the curriculum is designed by the Chinese Ministry of Education, which basically defines its content and assessment. The curricula that the Ministry of Education mandates include foreign languages, computer science, social sciences and the humanities. These curricula focus in large part on knowledge acquisition, and therefore the departments in charge of these particular courses seldom participate in PBL activities at ZU. In addition, the students are also required to pass the College English Test 4, College English Test 6 and the National Computer Test Level 2 to fulfill the basic requirements for graduation. These compulsory courses take up a large amount of the students' time and energy, and they represent a significant academic burden. Students have to prioritize the completion of the national curriculum and the tests over their PBL activities.

8.2.5 University Agenda

If the evaluation system for staff members and the assessment of students represent the tradition of the university, then the university agenda points to the current university practice. Fullan (2001) asserted that too many university agendas confuse organizational members in such a way that the university agenda may become destructive of staff and student motivation. As noted in the description of the XU case, the university introduced a number of agendas at the same time, which meant that staff members had to satisfy university requirements rather than being able to concentrate on PBL implementation. Moreover, some university agendas had already created a significant mismatch between what PBL needed and what the university wanted. Some staff members commented that it showed a lack of vision for the university to push for a new PBL approach (XU 2010 PBL model) when the first PBL generation (EE PBL and ACME PBL) had not been consolidated into the institution. What the university needs, they commented, is a deep analysis of the previous PBL model and a consolidation phase, rather than the abandonment of the previous PBL model. Another oft-repeated complaint is that the university's agenda of staff reduction is contradictory to PBL's need for more staff. The staff reduction policy engendered quite low morale. Thus staff motivation is rather theoretical than actually present, although the university does, in fact, claim to motivate staff members.

From the above discussion it can be seen that although the university made some technical changes to its organizational systems, such as developing a funding project, adding teaching elements in the teacher evaluation system and creating staff development programs, on the whole, these technical adjustments have not been powerful enough to motivate all of the organizational members to engage in PBL activities, as the dominant value of the organizational system remains contradictory to PBL implementation. The staff members at the university are directed to conduct increasing amounts of research rather than teaching; the updated staff evaluation policy seems to place a far greater emphasis on research. Staff members at the affiliated hospitals are more likely to engage in clinical work than PBL implementation due to the hospital's policy and their workload. The students at the university, although they seem to recognize the benefits of PBL, have a tendency to highlight the importance of knowledge acquisition and traditional learning methods due to the assessment methods used in their courses. Furthermore, the large number of national curriculum tests distracts students from participating in PBL activities. At XU, the university agenda means that staff members struggle to satisfy the managerial requirements that intrude somewhat into staff members' practice. As a result, the organizational members' motivation for engaging in PBL activities is generally low. In the following section, we are primarily concerned with dealing with the multifaceted nature of the modern university, which has a large impact on the motivation of the organizational members to participate in teaching activities (including PBL).

8.3 Research or Teaching?

The conventional views of curriculum innovation have the tendency to reduce the multifaceted nature of the university to an institution that only has teaching missions. In this simplified profile, the university's main responsibility is to provide a supportive learning environment for students and promote their learning outcomes. According to this perspective, changes in the curriculum only involve teaching- and learning-related elements. Biggs and Tang (2007) argue for the importance of aligning learning objectives, teaching activities and assessment methods when initiating curriculum change, implying that when one of these elements is changed during the curriculum innovation process, other elements should also be adjusted accordingly. Kolmos and De Graaff (2007) propose that a curriculum change should cover six elements: learning objective, teaching and learning activities, teachers, students and assessments. Both Kolmos (2001, 2002, and 2007) and Moesby (2004) contend that in order to conduct a full curriculum change such as a PBL implementation, changing the organizational culture is crucial. An essential element of the organizational culture is staff members' view of learning. Therefore, changing staff perception of learning is crucial for the success of the curriculum change. These perspectives on making curriculum changes can be categorized as the systematic curriculum change approach, as opposed to the traditional curriculum change approach, which only focuses on a single element such as teaching method, or assessment. These systematic views of curriculum change make sense to some extent. Nevertheless, the teaching/learning dimension of the university is significantly influenced by other, non-teaching dimensions of the university.

Here we should be aware that the nature of the university has already changed from a teaching institution aimed at preserving and transferring knowledge to students, as in "the good old days," to a complex organization in modern society. The complexity of the university as an organization is termed as loosely coupling systems (Cohen and March, 1974; Birnbaum, 1988). Furthermore, the multifaceted nature of the modern university is described as a multiversity (Kerr, 2001), comprising various departments, faculties, sectors, research centers, etc. that perform different functions and embody different values. The nature of the multiversity shapes the current university profile. Traditional universities "should have a soul, a single animating principle; the multiversity has several, some of them quite good, although there is much debate on which souls really deserve salvation" (Kerr, 2001: 9).

The complexity of the university in contemporary society poses huge challenges in dealing with different types of relationships. A major concern is how to keep the balance between research and teaching. There are much criticism that universities place too much emphasis on research over teaching (Page, 1990). The overemphasis on research at some research universities seems to be destructive to these universities' teaching function. Furthermore, various studies regarding the relationship between researching and learning have emerged (Martin and Berry, 1969; Faia, 1976; Jauch, 1976). From the cognitive point of view, research and teaching are mutually supportive and complementary (Paul II & Rubin, 1984; Schroeder, 1996). However, in terms of time and energy allocation, there are always tradeoffs between teaching and research (Jauch, 1976). Furthermore, although teaching and research are complementary, staff members at the university are in a process of "compensatory selection" between these two major activities as they move through their careers, which makes teaching and research less compatible (Faia, 1976).

As to the staff, within the context of organizational complexity, the role of staff at the university can be seen from various perspectives: they are instructors or tutors and are responsible for delivering the prescribed body of learning content to the students and are obligated to create a supportive learning environment for students; they are researchers carrying out scientific research activities and producing publications. In addition, due to the particularity of the context of the university, the staff at the university may perform other functions: they are also, for instance, medical doctors conducting clinical work at ZU. The staff seems to be trapped in a dilemma of teaching and research at higher education institutions. The overemphasis on research at universities has become a major challenge to the teaching since the staff members are forced to focus on the research rather than the teaching in order to fulfill the university's requirements for research such as publications, research funding, etc.

Many studies (Paul II & Rubin, 1984; Schroeder, 1996) suggest that research and teaching can complement each other, asserting that having both together makes sense for the cognitive process. Research leads to the production of knowledge, which can be applied to teaching, and research can cultivate habits of dealing with the unknown and self-inquiry that, in turn, may be useful in a teaching context such as PBL where the teachers are constantly challenged by the students. However, when considering the allocation of time and energy, the overemphasis on the research forces the staff to endeavor into research activities rather than teaching. For many staff members at the university, it is the research instead of the teaching that becomes their priority. The "publish or perish" atmosphere enormously distracts the staff from teaching activities to researching. However, for many staff members, conducting research has become routine, and they only see PBL as a temporary bureaucratic task. Once the task is complete, they return to their primary work of conducting research.

Role conflict seems to be quite common in large, complex organizations where organizational members are expected to perform different roles, some of which may even be contradictory (Gupta, & Jenkins, 1985; Cooper et al., 2001). Since the university is a complex organization performing various functions, the staff seems to be trapped in the dilemma of role conflict (e.g. Faia, 1976; Schroeder, 1996; Holbrook, Brodin, Balciuniene, et al., 2008; Vijava and Hemamalini, 2012). Thorough exploration of role conflict, its causes and its impact upon the organization is far beyond the scope of this study. Here, I am particularly concerned with the relationship between role conflict and PBL implementation at institutions of higher education. It is worth noting that whether the staff is motivated to conduct teaching activities (PBL activities included) is significantly dependent on whether their roles as teachers are emphasized by the university in practice. On the basis of previous descriptions, we can see that salary and promotion policy favor research rather than teaching at ZU. If this trend continues, the entire teaching enterprise at the university is likely to be damaged, not to mention the implementation of PBL. Administratively, evaluations tend to influence the direction that the staff chooses to follow. If teaching is to enjoy a status equal with research, then the evaluation system must be changed to account for time input, and significant rewards for teaching must be offered (Jauch, 1976). Otherwise, the intended impact of any new educational innovation to promote teaching and learning will be undermined. To solve this problem, we should carefully reflect on the role of the modern university. Is it an institution for cultivating students, or an institution for carrying out other tasks (such as research, social service, etc.)? PBL cannot be widely respected and accepted if the university emphasizes research rather than teaching.

Some staff members insist that lack of resources is a significant challenge for PBL implementation at a university. This is partially true, since PBL requires more staff members, group rooms, opportunities to access library resources and so on. Without sufficient resource support, PBL implementation cannot progress smoothly. More importantly, it is the problematic distribution of resources favoring non-PBL agendas, rather than the total lack of resources, that hinders PBL implementation at the university. If the resources are distributed to non-PBL activities, such as research or clinical tasks, staff members at the university are more likely to engage in these non-teaching activities rather than in PBL.

9

Management of the Change to PBL

The stories of XU and ZU demonstrate a diversity of interpretations regarding what PBL is, what group work is and how staff members and students are motivated in the change process. The stories, meanwhile, show the consequences of the diverse interpretations of PBL issues. For example, XU attempts to regulate organizational members' behavior by setting up a common PBL model throughout the organization. However, due to the diversity of PBL interpretations, the managerial intention of implementing a single PBL approach seems to have created notable tensions at the university. At ZU, there seems to be little tension, since the university encourages individual exploration and creation. However, the situation seems a bit chaotic; there is very little coordination between different staff members. The problem thus occurs: as educational theorists or practitioners, how should we deal with this diversity?

Managers employed a normative approach when making the transition to PBL at XU and ZU. By normative we mean that the managers developed a goal for the organizational change and afterwards adopted various techniques such as economic stimulus and educative strategy to change the values and actions of organizational members in accordance with the managerial vision. The normative approach implies that there should be a correct, effective or appropriate objective process with which the organizational members have to comply. It also suggests that any thoughts and behaviors deviating from managerial expectations need to be examined or even eliminated. We can see that the normative approach was used in different manners at ZU and XU, and eventually led to different results. At XU, the university management attempted to impose a managerial interpretation of PBL throughout the entire organization, which afterwards resulted in notable tension due to the existence of various PBL interpretations. At ZU, the senior managers encouraged different departments and individuals to make innovations regarding PBL and, as a result, variations of PBL interpretations developed that created a relatively chaotic situation. Therefore, the extent to which the normative approach should be used should be examined.

9.1 Limitations of Normative Approaches

A normative approach involves the guidance and control of organizational members' actions and attitudes. The senior manager at the university is willing to see the departments, the staff members and the students, change their behavior in accordance with managerial guidance and expectations. The organizational members at the bottom level are not only required to transform their behavior, but also to change their values and concepts. A variety of techniques, such as staff development programs, project funding and evaluation and assessment adjustments are consequently deployed to interfere with the organizational members' actions and values. These managerial attempts have partially succeeded at both XU and ZU, since a number of staff members and students are willing to accept the idea of PBL and become engaged with PBL activities. However, we should carefully reflect on the normative approach and examine its advantages and drawbacks.

9.1.1 Creating Tensions

Since organizational members have their own interpretations of PBL, which are not always in keeping with managerial efforts, the use of a normative approach for PBL implementation may sometimes create great tension in the organization. This may in turn undermine the transition process towards PBL. It seems that for the staff members who claimed to have an opinion of PBL, the managerial efforts were huge interruptions for them. For the staff members who believed in the value of the traditional educational approach, PBL challenged their own conceptions of education and made them feel insecure and uncomfortable. Therefore, these staff members were reluctant to participate in PBL-related activities. As we saw in the case of XU, there was a significant level of staff resistance to PBL, which was regarded by many other staff members as a huge challenge for PBL implementation.

9.1.2 Incomplete Control

It is clear that organizational members' values and actions are not always consistent with the managerial vision, even though the management level uses normative approaches to guide staff members' actions and seeks to have an impact on staff members' thoughts. In other words, complete control of the organization and the change process is almost impossible, despite the use of well-designed change plans. At XU, although the managerial interpretation of PBL was imposed throughout the entire university, variations in PBL interpretations still existed at both departmental and individual levels and were not always consistent with the managerial interpretation. Some interpretations of PBL, such as the one equating PBL to problem-solving learning, even contradict what was proposed by the university, and are likely to continue to persist regardless of managerial attempts to impose a single PBL interpretation throughout the university. At ZU, the social learning aspect of PBL is interpreted in a different manner from its theoretical ideal, which aims to promote the interaction among the students and that between the students and the teachers. Both staff members and students developed alternative strategies to enhance the effectiveness of the group and the class, though they claimed social learning had been included in their teaching and learning process. Furthermore, both XU and ZU offered a large amount of institutional support to motivate the staff to participate in PBL activities, but the institutional support did not necessarily guarantee that the staff members and students would fully engage in PBL activities.

The failure to gain complete control of the organization is partially accounted for by the fact that the organization is, first and foremost, a dynamic and complex system with a high level of uncertainty, as is inferred by anarchic organization proponents (Cohen, March and Olsen, 1972). The organizational process cannot be fully controlled due to the obscure nature of organizational objectives, the flow of organizational members, the uncertainty of the problems encountered by the organization, etc. Furthermore, unlike other organizations, such as business firms, a university demonstrates a "unique dualism in organizational structure" in its administration (Corson, 1960, p.43), which means that the power structure at a university is highly fragmented and different groups or parties retain their own "spheres of influence" according to their professional concerns. The staff claims the primary influence over issues such as curriculum and degree requirements, whereas managers claim matters such as budgets, long-term planning and construction of programs. Thus managers cannot interfere with all aspects of the university system, no matter how powerful they are nor what kind of strategy they employ to manage the university. Although managers can implement a systematic change throughout the organization, they are unlikely to completely control what happens in the classroom, since the classroom belongs to the teachers' sphere of influence. How teaching and learning works in a classroom is mainly dependent upon the individual teacher and the students, not the administrative managers. Moreover, organizational change involves the subjective participation of organizational members, which cannot be readily manipulated by technical means. Organizational members' interpretations of a particular phenomenon are significantly conditioned by their "fore-structure" (Gallagher, 1992; Gadamer, 1995), which largely comprises their tradition and personal experience rather than technical means. As this research shows, the diversity of PBL interpretations can be accounted for by the difference in organizational members' work experience and educational experience. At XU, the staff interpretation of PBL as synonymous with problem solving can be linked to the staff members' working experience. At ZU, the students' emphasis on knowledge retention can be connected to their primary and secondary school experience.

9.2 Productivity and New Possibilities

A normative approach is intended to guide the organizational members' values and actions, and to remove and eliminate those that are not consistent with the managerial interpretation and vision. Though it is beneficial in bringing about an effective change process, it may be destructive to the organization's ability to innovate. Here we argue for a new perspective on variations in PBL interpretations. The conventional wisdom tends to view these variations as incorrect or distorted perspectives that need to be controlled, converted or even eliminated. They are usually considered to be detrimental to the change process, since they sometimes disturb, distort or obstruct the change process. However, if taken positively, they may disclose some prospective and promising possibilities that transcend our rational design and imagination. Thus we should not readily deny the value of practices and perspectives that may seem to be inconsistent with the planned ones, since they are likely to open up the collective mindset to new possibilities. That is, we should treat these variations in a tolerant and constructive manner, assuming them to be productive.

We do not use the term "productive" in the sense of enhancing effectiveness or harvesting expected outcomes, but in the sense of constantly making something relevant to oneself, creating new possibilities and opportunities and thus renewing tradition (Gallagher, 1992). The productivity of these unintended PBL interpretations can be regarded as attempts to localize the notion of PBL in a specific context, because departments and individuals always bring their own traits and experiences to the practice of teaching when they interpret PBL.

The implementation of PBL at the departmental level can be regarded as an interpretative practice at the local level. As has been pointed out, all interpretative practices are local. In this sense, the preceding problematic PBL approaches can be considered as the staff members' interpretation of PBL out of its specific context. As a staff member at ZU commented:

> "Chinese students are getting accustomed to following the teachers' instructions during the educational process. They don't know how to learn by themselves. . . .The European students and American students have self-teaching abilities. If you leave them alone, they have no problem. But if you let Chinese students study alone, they will become a total mess. . . .The moderate or the transition method is to give them more opportunities to study by themselves while offering them strong guidance rather than simply leaving them alone."

From this quotation, we see that strong guidance in PBL, although not quite consistent with the principles of PBL, fits the Chinese context, since Chinese students traditionally rely upon the instruction of their teachers. In addition, the students who are in charge of organizing group activities or class discussions at ZU may also benefit from their PBL experience, as a staff member maintained:

> "The student giving lectures to fellow students is not just lecturing; he may, on occasion, organize and facilitate class discussions, depending on the situation. Though not perfect, he is still able to practice his communication skills and management skills. Since all of us have the chance to present and host the discussion, there are some opportunities for us to practice our PBL skills. It is a pity that not everyone can stand on the platform and give presentations, but what else can we do? We have so many students. The academic burden is pretty heavy."

As we saw earlier, some PBL interpretations, though they seem to deviate from the ideal of PBL theories, can be viewed as a compromise between the theoretical ideal and the practical conditions. The compromise may imply new possibilities in the future. For instance, when the entire organizational system tends to favor knowledge-oriented learning, it is a great step towards implementing the "real" PBL when a university introduces PBL even as a way of addressing major social or disciplinary issues, or executes the blended use of PBL and LBL.

9.3 Necessity of Normative Approaches

A normative approach is necessary for any organizational change because it is beneficial in keeping the organizational members' values and actions in accordance with the managerial vision, and thus assuring the effectiveness of the changeover. The lack of a normative dimension may result in a relatively chaotic situation, as we saw at ZU. At ZU, due to managerial attempts to encourage individual innovation, and due to the lack of leadership, the departments and staff developed a variety of PBL forms. Although PBL at ZU appeared to be prospering, some interpretations of PBL were in essentials far from the core values of PBL. For example, minimizing the interaction among students so as to maintain group effectiveness is inconsistent with PBL rationale, and makes students passive learners, just as they are in traditional education contexts. Therefore, we should be aware of the influence of these PBL interpretations on students' learning as well as on the overall organizational change plan. Furthermore, since there was a lack of guidance regarding the schedule of the PBL curricula, it resulted in a chaotic schedule of PBL tasks across different courses. At times, a series of PBL courses might coincide and would create an enormous academic burden for the students, whereas during some periods, there might be no PBL at all. Given this problem, a senior staff member at ZU contended:

> "Systematic design is quite important because it can coordinate the actions of different departments that are now just doing PBL on their own. Also, it can make PBL implementation more orderly. There should be an overall plan in future years that delineates what are we going to achieve, how much PBL work the students should be doing, etc."

Thus we argue that it is necessary for the university to offer more managerial support for PBL.

9.4 Revisit the Use of Normative Approaches

From the above discussion we can see that the university has a dilemma regarding the use of normative approaches when making the change towards PBL. On the one hand, a normative approach is necessary for PBL implementation since it can potentially align the organizational members' values and actions with the managerial vision and thus enhance the effectiveness of the organizational process. The lack of a normative dimension is likely to lead to a chaotic situation, which might hamper PBL implementation, as seen in the ZU case. However, overemphasis on the normative approach may cause tension within the organization, as we saw in the XU case, where significant tension emerged when the university imposed the managerial interpretation of

PBL throughout the organization. Furthermore, removing and eliminating the seemingly fallacious PBL interpretations and innovations may discount their potential strengths and possibilities. Therefore, the university needs to strike a balance between standardizing the organization and giving staff members more freedom to explore and innovate, which poses significant challenges to the organization in the change process. To standardize means to maintain the effectiveness of the organization by aligning the organizational members' actions with the managerial vision. The opposite of the normative approach is to use a more flexible method and give the organizational members more space to innovate. However, there is no straight answer to how the organization should strike the balance between guaranteeing organizational effectiveness through the use of normative approaches and staying open to new possibilities by providing more freedom to organizational members, as each choice has pros and cons. Seeking for balance between the choice of greater effectiveness and greater flexibility continues to be a significant challenge to the university in the process of change to PBL.

10

Concluding Remarks

10.1 Summaries of This Book

During the past half-century, PBL has been adopted by many institutions of higher education as an alternative educational approach because it can bring considerable benefits, including increasing students' learning motivation and cultivating students' soft skills, and also because it represents a new way of perceiving knowledge and learning. However, the educational change towards PBL is a significant challenge for institutions of higher education that have a long history of using the traditional educational approach. Therefore, it is necessary to explore the challenges and obstacles during the change to PBL in institutions of higher education. Since current studies regarding the challenges for PBL implementation lack a systematic perspective and a consideration of the organizational members' understanding of PBL implementation, this study is also aimed at filling some of the gaps of contemporary research while addressing the challenges and obstacles of the transition process.

In the two stories of the organizational change to PBL, various types of PBL interpretations emerged during the change process. They can be roughly categorized into three distinct levels: managerial interpretations, departmental interpretations and individual interpretations. They differ significantly from each other regarding the correct proportion of the PBL component in the whole curriculum system, how to coordinate PBL with LBL, what the 'P' in PBL means and so on. Some of the interpretations are consistent with the ideal form of PBL, while others seem to be a far cry from PBL theories. Due to the fact that different change strategies were used at each university, different interpretations of PBL led to different results: at XU, tensions occurred as the managers attempted to impose a single PBL definition throughout the entire university, while at ZU, the managerial encouragement of personal/departmental innovation resulted in a situation of chaos to some extent.

At the micro level, the way in which social learning occurs in the classroom needs more research attention, since it has rarely been discussed before. Specifically, I am interested in the scale of interaction among students, peer learners in the same group and their tutors. In this study, though some groups enjoyed high levels of interaction and communication, many groups and individuals adopted the role distribution strategy so as to enhance their group effectiveness, which largely confines the group dynamics. In some groups, the interaction among group members seemed to be at a minimal level. In a large class, the intended discussion effectively became a traditional lecture where the majority of the students became silent learners.

In addition to the diversity of the interpretations of PBL and social learning, several major flaws can be noted. The first one is that most interpretations of PBL place their emphasis on designing (both curriculum and teaching) rather than learning. Furthermore, there is a strong focus on knowledge retention instead of the growth of the learners. In addition, the learning outcome is heavily emphasized, indicating a relative ignorance of the learning process, which should be what distinguishes PBL from LBL. Furthermore, PBL becomes widely reduced to an instrumental technique which is only advantageous for its instrumental merits. This cognitive reduction of PBL may be destructive to the conceptual change of the organizational members regarding learning and knowledge, since PBL represents an alternative way of learning and knowing which is far more than simply some instrumental benefits. These common flaws in the interpretation of PBL and social learning call for a radical conceptual change in learning and knowledge for organizational members at the university. Without it, a successful change to PBL is unlikely to happen. For example, a large number of the staff members and students at ZU, although claiming to be conducting PBL, tend to equate learning with the acquisition of knowledge, which is further reduced to the memorization of information. Clearly, conceptual change for PBL is not an easy task.

Although many researchers advocate a systematic perspective (e.g., Biggs, John and Tang, 2007; Kolmos and De Graaff, 2007) when dealing with curriculum change, they have not addressed the multifaceted nature of a modern university. Universities in the contemporary world undertake various roles that pose enormous challenges to organizational members at the university. When linking this to the motivation issue, it can be seen that many university agendas, through the means of salary and promotion policies, largely shift staff motivation from teaching to non-teaching activities such as research, clinical work, etc. If this trend continues, any educational innovation attempt to enhance teaching and learning quality, including PBL implementation, is likely to be undermined. As the evaluation and promotion systems have a strong influence in directing staff members' behavior, there should be scrutiny on these systems when introducing curriculum change. Most importantly, there should be a reflection on the role of the university and the staff members within it. It is quite difficult for a university that makes research its priority to become interested and engaged in a teaching mission. If the staff members in such a university are not enthusiastic about teaching activities, they are very unlikely to become interested in teaching innovations, such as PBL.

Finally, with regard to the management of the change, the university has to strike a balance between the use of a normative approach and giving the organizational members enough freedom to make their own explorations and innovations. A normative approach is necessary for the organizational change since it directs the organizational members' values and actions in accordance with the managerial vision, and this approach is therefore beneficial for the organization in carrying out an effective change. However, the normative approach emphasizes a degree of control of the organizational members' values and actions, and causes tensions in the organization, as seen in the case of XU. Furthermore, the use of the normative approach may neglect the value of certain PBL interpretations, deeming them incorrect without considering the new possibilities they may add to the current PBL wisdom. Therefore, we should exercise caution in the use of normative approaches. Giving freedom to organizational members is quite valuable because it can stimulate organizational members to make innovations and create new possibilities. However, the lack of the normative dimension may lead to a chaotic situation, as seen at ZU. Therefore, there is no straightforward answer to whether a university should lean towards the use of normative approaches or allow more freedom for organizational members. The balance is highly dependent on the organization's preference for effectiveness (more control) or flexibility (more freedom). Perhaps the real answer is that the organization is not making a choice between good and bad, but between bad and worse.

When we are making organizational change, the rational approach (identifying objectives, selecting the strategy, motivating organizational members, etc.) is usually based upon the assumption that there is a perfect way (most effective, optimistic, appropriate) to implement a change. By revealing the dilemma that the organization is facing (organizational effectiveness vs. new possibilities) we attempt to refute the widely accepted assumption of a perfect change approach, since any approach has its own strengths and limitations. A strong normative approach could secure organizational effectiveness to some extent, but it might create internal tensions within the organization. Encouraging individual explorations could create new possibilities, but it might result in organizational chaos. Therefore, what matters most is not to seek the best change approach, but rather to be aware of both the strengths and the disadvantages of each approach.

10.2 Reflections on This Study

The focal point of this study is to address the challenges and obstacles for institutions of higher education that are in the process of changing to PBL. In addition, this study focuses on organizational members' subjective world in order to add value to some of the current studies dealing with the challenges and obstacles to the organizational change to PBL as explained in Chapter 1. Therefore, this study can contribute to the current knowledge field and to educational practices in the following respects:

First, this study consolidates and further develops the arguments that systematic perspectives (e.g. Biggs, John & Tang, 2007; Kolmos& De Graaff, 2007) are greatly needed when dealing with the organizational change to PBL. We argue that when dealing with PBL implementation it is not sufficient to simply change a single element of the curriculum, nor is it enough to change only the elements relevant to teaching and learning (e.g. Biggs, John & Tang, 2007). Rather, we need to reflect on the multifaceted nature of modern universities and the role of the university in contemporary society.

In addition, this study explores organizational members' understanding and interpretations of PBL implementation, a phenomenon that is almost entirely missing from the previous research regarding the organizational change to PBL. By exploring organizational members' perceptions of PBL implementation, this study is able to show how PBL is actually delivered in reality, rather than how it is designed, thus enabling us to achieve a deep understanding of the challenges and obstacles for the organizational change to PBL.

Furthermore, this study offers a rich description of the stories of two universities that are in the process of transforming their educational paradigms to PBL. The rich description demonstrates the dynamics and complexity of the process of organizational change to PBL, and also addresses the challenges and obstacles to it by revealing the causal relationships between the organizational members' interpretations of PBL implementation (including PBL, social learning and motivation) and their influence on the organization.

At a practical level, although this study is not attempting to formulate guidelines for PBL implementation, we can nevertheless argue that this study can assist educational practitioners such as university managers and staff members to critically examine and reflect on their daily thoughts and practices. This exercise can help them to become "*reflective practitioners*" (Schön, 1983) who can constantly examine and improve their practices by referring to the evidence and discussions regarding PBL in this study. For example, managers at universities need to realize the importance of conceptual changes for the organizational members when introducing educational innovations and provide more institutional support to make these changes happen. They also need to endeavor to reflect on their change strategies and strike a balance between maintaining the effectiveness of the organizational change and offering staff members more opportunities for exploration and creating more possibilities. This study may assist staff members to reflect on their opinions and practice of PBL. For example, it is important to examine the tendency to equate PBL with problem-solving learning, case studies or a curriculum-design approach, or simply to view it as an instrument to achieve the realization of educational values. Such notions may undermine the implementation of PBL.

Due to various practical factors, this study is not free from flaws. When dealing with PBL implementation, Savin-Baden and Major (2004) suggest that the cultural context should be taken into consideration, since it tends to influence PBL implementation. Therefore, it is necessary to examine the relationship between a particular cultural context and PBL implementation in the future. Furthermore, when a higher education institution initiates an organizational change towards PBL, the disciplinary culture and its impact on PBL implementation should also be taken into account. In addition, currently, the research on PBL implementation focuses primarily on the design of PBL curricula and activities rather than on how PBL actually performs in reality. Therefore, it is necessary to examine how PBL actually performs in practice, especially at the micro level. Since PBL is concerned with the learning of students rather than the teaching of teachers, it is particularly essential to address how students perform within PBL environments so as to understand what happens to the students' conception of learning, and learning habits generally within PBL environments. Although this study addresses how students act in PBL groups and classes to some extent, the examination of how PBL performs at the micro level is far from sufficient. The limitations of this study point to the next steps for future research.

References

1. Alarcao, Isabel (2007). Changing to Project-based Learning: the Role of Institutional Leadership and Faculty Development. In Kolmos, A. and De Graaff, Erik (Eds.). Management of Change. Rotterdam: Sense Publishers. 69–82.
2. Albanese, M. (2000). Problem-based Learning: Why Curricula Are Likely to Show Little Effect on Knowledge and Clinical Skills. Medical Education. 34, 729–738.
3. Albanese, M. and Mitchell, S. (1993). Problem-based Learning: A Review of the Literature on its Outcomes and Implementation Issues. Academic Medicine, 68, 52–81.
4. Armstrong, E. (1997). A Hybrid Model of Problem Based Learning. In Boud, David and Feletti, Grahame (Eds.). The Challenge of Problem-Based Learning (pp.136–150). London: Kogan Page.
5. Ausubel, D. P., Robinson, F. G. (1969). School Learning: An Introduction to Educational Psychology. New York: Holt, Rinehart & Winston.
6. Bair, C. R. (1999). Meta-synthesis: a New Research Methodology. Paper presented at the Annual Meeting of the Association for the Study of Higher Education, November 18–21, Santonio, Texas, 26p (ERIC document NO. ED473 866).
7. Baldridge, J., Curtis, D., Ecker, G. & Riley, G. (1978). Policy Making and Effective Leadership. San Francisco: Jossey-Bass Publishers.
8. Baldridge, J. V. (1971). Power and Conflict in the University: Research in the Sociology of Complex Organizations. New York: John Wiley and Sons.
9. Bandura, Albert (1986). Social Foundations of Thought and Action: A Social Cognitive Theory. Englewood Cliffs, NJ: Prentice-Hall.
10. Barge, Scott (2010). Principles of Problem and Project Based Learning: The Aalborg PBL Model. Retrived from: http://www.google.dk/url?sa=t&rct=j&q=barge%20aau%20pbl&source=web&cd=5&cad=rja&ved=0CDUQFjAE&url=http%3A%2F%2Ffiles.portal.aau.dk%2Ffilesharing%2Fdownload%3Ffilename%3Daau%2 Faau%2F2010%2F~%2Fpub%2FPBL_aalborg_modellen.pdf&ei=92yEULfNArOK4gTk3IHgBg&usg=AFQjCNE17kJLxc3w6mOIvKESyNE-Xv2dww.

11. Barr, Robert B. and Tagg, John. (1995). From Teaching to Learning: A New Paradigm for Undergraduate Education. Change, 27(6): 12-26.
12. Barrett, T. and Moore, S. (2011). An Introduction to Problem-based Learning. In Barrett, Terry and Moore, Sarah (eds.). New Approaches to Problem-Based Learning: Revitalising Your Practice in Higher Education. New York: Routledge.
13. Barrows, H. S. and Tamblyn, R. M. (1980). Problem-Based Learning: An Approach to Medical Education. New York: Springer.
14. Barrows, H. S. (1986). A Taxonomy of Problem-based Learning Methods. Medical Education, 20, 481–486.
15. Barrows, H. S. (2002). Is It Truly Possible to Have such a Thing as dPBL? Distance Education, 23(1), 119–122.
16. Baum, Joel A. C. (1996). Organizational Ecology. In Clegg, Stewart R., Hardy, Cynthia and Nord, Walter R. (Eds.). Handbook of Organization Studies. Thousand Oaks, CA: Sage.
17. Bauman, Z. (1998). Globalization: the Human Consequences. Cambridge: Polity Press.
18. Bauman, Z. (2000). Liquid Modernity. Cambridge: Polity Press.
19. Bauman, Z. (2005). Education in Liquid Modernity. The Review of Education, Pedagogy and Cultural Studies, 27, 303–317.
20. Biggs, John and Tang, Catherine (2007). Teaching for Quality Learning at University (3rd edition). Maidenhead: Open University Press.
21. Birnbaum, R. (1988). How Colleges Work: The Cybernetics of Academic Organization and Leadership. San Francisco: Jossey-Bass.
22. Blumberg, P. and Michael, J. (1992). Development of Self-directed Learning Behaviors in a Particially Teacher Directed Problem-based Learning Curriculum. Teaching and Learning in Medicine, 4(1), 3–8.
23. Bogdan, R and Biklen, S. (1992). Qualitative Research for Education (2nd edition). Boston, MA: Allyn and Bacon.
24. Boud, D. (1985). Problem-Based Learning in Education for the Professions. Sydney: Higher Education Research and Development Society of Australia.
25. Boud, D. (Eds.) (1985). Problem Based Learning in Education for Professions. Sydney: Higher Education Research and Development Society of Australasia.
26. Bowe, Brian (2007). Managing the Change from Traditional Teaching to Problem-based Learning in Physics Education. In Kolmos, A. and De Graaff, Erik (Eds.). Management of Change. Rotterdam: Sense Publishers. 83–91.

27. Bowe, Brian and Cowan, John (2004). A Comparative Evaluation of Problem Based Learning in Physics: A Lecture Based Course and A Problem Based Course. In Savin-Baden, Maggi (Eds.). Challenging Research in Problem Based Learning (pp. 161–173). Berkshire: McGrawHill Education.
28. Brown, John and Duguid, Paul (1991). Organizational Learning and Communities of Practice: Toward a Unified View of Working, Learning, and Innovation. Organization Science, Vol.2, No. 1, 40–57.
29. Cao, Z., Gabb, R.(2006). Patterns of Progress and Attrition in Commencing Higher Education Students at VU. Melbourne: Victoria University, Post-compulsory Education Centre.
30. Chin, R., and Benne, K. D. (1985). General Strategies for Effecting Changes in Human Systems. In Bennis, W. G., Benne, K. D. and Chin, R. (Eds.). The Planning of Change (4th edition). New York: Holt, Rinehart and Winston.
31. China Medical University (2010). Organizational Structure. Retrieved from http://www.cmu.edu.cn/new/showpage.asp?pageid=333.
32. China Medical University. (2004). The Notification of Funding Project Regarding Problem Based Learning at China Medical University (Chinese original version: Guan Yu Zhong Guo Yi Ke Da Xue Yi Wen Ti Wei Ji Chu De Xue Xi Jiao Xue Gai Ge Xiang Mu Li Xiang De Tong Zhi).
33. China Medical University (2010 a). Retrieved from http://www.cmu.edu.cn/new/showpage.asp?pageid=314.
34. China Medical University (2010 b). Organizational Structure. Retrived from http://www.cmu.edu.cn/new/showpage.asp?pageid=333.
35. Cohen, Dan (2005). The Heart of Change Field Guide: Tools and Tactics for Leading Change in Your Organization. Massachusetts: Harvard Business School Press.
36. Cohen, L. Manion, L. and Morrison, K. (2007). Research Methods in Education (6th edition). New York: Routledge.
37. Cohen, Michael D. & March, James G (1974). Leadership and Ambiguity. Boston, Massachusetts: Harvard Business School Press.
38. Cohen, Michael D., March, James G. and Olsen, Johan P. (1972). A Garbage Can Model of Organizational Choice. Administrative Science Quarterly, 17(1), 1–25.
39. Colbeck, Carol L (2002). Balancing Teaching with Other Responsibilities: Integrating Roles or Feeding Alligators. Paper presented at the Annual Meeting of the American Educational Research Association (New Orleans, LA, April 1–5, 2002).

40. Collaboration and Structural Reform Fund (2006). Projects Funded through the 2006 Funding Round. Retrieved from http://www.dest.gov.au /NR/rdonlyres/2F19DCD0–9B06–4CF2–8367–5FC8E112FB39/15971/ Casr_2006_funded.pdf.
41. Colliver, Jerry (2000). Effectiveness of Problem-based Learning Curricula: Research and Theory. Academic medicine, Vol.75, No.3, 259–266.
42. Colvin, D. L. (2005). Policymaking and Governance in a Multi-campus State University System: the Case of General Education Reform at the State University of New York. Unpublished Ph. D Dissertation. The University of Pennsylvania.
43. Commonwealth of Australia (2006). Higher Education Support Act 2003. OTHER GRANTS GUIDELINES 2006.
44. Connolly, D. and Silen, D. (2011). Empowering Tutors: Strategies for Inspired and Effective Facilitations of PBL Learning. In Barrett, Terry and Moore, Sarah (Eds.). New Approaches to Problem-Based Learning: Revitalising Your Practice in Higher Education. New York: Routledge.
45. Cooper, C. L., Dewe, P. J. and O'Driscoll, M. P. (2001). Organizational Stress: A Review and Critique of Theory, research, and applications. Thousand Oaks, California: Sage Publications.
46. Corson, J. J. (1960). Governance of Colleges and Universities. New York: McGraw-Hill.
47. Creswell, John (2009). Research Design: Qualitative, Quantitative, and Mixed Methods Approaches. London: Sage.
48. Curtis, S., Gesler, W., Gesler, S. and Washburn, S. (2000). Approaches to Sampling and Case Selection in Qualitative Research: Examples in the Geography of Health. Social Science and Medicine, (50): 1001–1014.
49. Czarniawska, Barbara. (1997). Narrating the Organization: Dramas of Institutional Identity. Chicago: The University of Chicago Press.
50. Czarniawska, Barbara. (1998). A Narrative Approach to Organization Studies. Qualitative Research Methods Series 43. London: Sage.
51. Daft, Richard and Weick, Karl (1984). Toward a Model of Organizations as Interpretive Systems. The Academy of Management Review. Vol.9, No. 2, 284–295.
52. Dalin, Per (2004). School Development: Theories and Strategies. London: Continuum International Publishing.
53. De Graaff, Erik and Kolmos, Anette (2003). Characteristics of Problem-based Learning. International Journal of Engineering Education, 19(5), 657–662.

54. De Graaff, Erik, and Cowdroy, Rob (1997). Theory and Practice of Educational Innovation Introduction of Problem-based Learning in Architecture: Two Case Studies. International Journal of Engineering Education, 13(3), 166–174.
55. De Oliveira, Jose Manuel Nunes (2007). Project-based Learning in Engineering: the Agueda Experience. In Kolmos, A. and De Graaff, Erik (Eds.). Management of Change. Rotterdam: Sense Publishers. 169–180.
56. De Vries, M., Schmidt, H. G. and de Graaff, E. (1989). Dutch Comparisons: Cognitive and Motivational Effects of Problem Based Learning on Medical Students. In Schmidt, H. G., Lipkin, M., De Vries, M., and Greep, J. M. (Eds.). New Directions for Medical Education: Problem Based Learning and Community Oriented Medical Education. New York: Springer-Verlag, 230–240.
57. DEST (2002). A Report by the Australia Chamber of Commerce and Industry and the Business Council of Australia for the Department of Education, Science and Training, Canberra.
58. DETYA (1998). Learning for life: review of higher education financing and policy, Higher Education Financing and Policy Reviewing Committee, known as the West Review, Department of Education, Science, Training and Youth Affairs, Canberra.
59. Dewey, John (1916). Democracy and Education: An Introduction to Philosophy of Education. New York: Macmillan (1966 paperback edition).
60. Dewey, John (1938). Experience and Education. Indianapolis: Kappa Delta Pi. Retrieved from http://books.google.dk/books?id=UE2EusaU53IC&pg=PA42&lpg=PA42&dq=whatever+conditions+interact+with+personal&source=bl&ots=EE-W-rLONh&sig=R6QLDjvL4J8S5kOjpdSVnxOXoMU&hl=en&sa=X&ei=SbYeUO7WIsbP0QXv54CgCw&ved=0CCoQ6AEwAA#v=onepage&q=whatever%20conditions%20interact%20with%20personal&f=false.
61. Dicicco-Bloom, B. and Crabtree, B. (2006). The Qualitative Research Interview. Medical Education, 40, 314–321.
62. Dochy, F., Segers, M., Bossche, P., Gijbels, D. (2000). Effects of Problem-based Learning: A Meta-analysis. Learning and Instruction, 2003(13), 533–568.
63. Dolmans D. and Schmidt, H. (1996). The Advantages of Problem Based Curricula. Postgrad. Med. J. 72, 535–538.
64. Donner, R. S. and Bickley, H. (1993). Problem Based Learning in American Medical Education: An Overview. Bulletin of Medical Library Association, 81, 294–298.

65. Dooley, Kevin (1997). A Complex Adaptive Systems Models of Organization Change. Nonlinear Dynamics, Psychology and Life Science. Vol.1, No.1, 69–97.
66. Duch, Barbara (2001). Models for Problem Based Instruction in Undergraduate Courses. In Duch, Barbara J., Groh, Susan E. and Allen, Deborah E. (Eds.) The Power of Problem-Based Learning (pp.39–46). Sterling: Stylus Publishing.
67. Faia, Michael A. (1976). Teaching and Research: Rapport or Mesalliance. Research in Higher Education. Vol. 4, No. 3, 235–246.
68. Fischler, H. (1994). Concerning the Difference between Intention and Actions: Teachers' Conceptions and Actions in Physics Teaching. In I. Carlgren, G. Handal & S. Vaage, S. (Eds.). Teacher's Minds and Actions: Research on Teachers' thinking and practice (pp. 165–180). London: The Falmer Press.
69. Flyvbjerg, Bent (2001). Making Social Science Matter: Why Social Inquiry Fails and How It Can Succeed Again. Cambridge: Cambridge University Press.
70. Flyvbjerg, Bent (2006). Five Misunderstandings about Case-study Research. Qualitative Inquiry, 12(2), 219–245.
71. Fossum, Lynn (1989). Understanding Organizational Change: Converting Theory into Practice. Boston, MA: Course Technology Crisp.
72. Freire, P. (1970). Pedagogy of the Oppressed. New York: Continuum.
73. Fullan, Michael (2001). The New Meaning of Educational Change. New York: Teachers College Press.
74. Gabb, R. (2006). 2004 First Year Experience Survey: Findings from Victoria University Respondents. Melbourne, Victoria University, Postcompulsory Education Centre.
75. Gabb. G., Milne, M., and Cao, Z. (2006). Understanding Attrition and Improving Transition: A Review of Recent Literature. Melbourne, Victoria University, Postcompulsory Education Centre.
76. Gadamer, Hans-Georg. (1995). Truth and Method (2nd revised edition). New York: Continuum (original work published in 1960).
77. Gallagher, Shaun (1992). Hermeneutics and Education. SUNY Press.
78. Gibbs, G. R. (2007). Analyzing Qualitative Data. In U. Flick (Eds.). The Sage Qualitative Research Kit. London: Sage Publications.
79. Golafshani, N. (2003). Understanding Reliability and Validity in Qualitative Research. The Qualitative Report, 8(4), 597–606. Retrieved from http://www.nova.edu/ssss/QR/QR8–4/golafshani.pdf.

80. Guba, E. and Lincoln, Y. (1985). Naturalistic Inquiry. Newbury Park, CA: Sage.
81. Guba, E. and Lincoln, Y. (1994). Competing Paradigms in Qualitative Research. In N. Denzin and Y. Lincoln (Eds.). Handbook of Qualitative Research (pp. 105–117). California: Sage.
82. Gupta, N. & Jenkins, G. D. (1985). Dual-career Couples: Stress, Stressors, Strains and Strategies. In T. A. Beehr & R. S. Bhagat (Eds.). Human Stress and Cognition in Organizations: an Integrated Perspective. New York: Wiley, 141–175.
83. Habermas, J. (1984). The Theory of Communicative Action, Vol. 1: Reason and the Rationalization of Society. Translated by T. McCarthy. Boston, MA: Beacon Press.
84. Hall, Richard (1999). Organizations: Structures, Processes and Outcomes. Englewood Cliffs, NJ: Prentice Hall.
85. Healy, M. and Perry, C. (2000). Comprehensive Criteria to Judge Validity and Reliability of Qualitative Research within the Realism Paradigm. Qualitative Market Research, 3(3), 118–126.
86. Henriksen, Lars Bo et al. (2004). Dimension of Change: Conceptualizing Reality in Organizational Research. Copenhagen: Copenhagen Business School Press.
87. Holbrook, W., Brodin, P. Balciuniene, I. et al. (2008). Balancing the Role of the Dental School in Teaching, Research and Patient Care. European Journal of Dental Education. 12 (suppl. 1), 161–166.
88. Huber, G. P. (1991). Organizational Learning: the Contributing Processes and the Literature. Organization Science. 2(1), 88–115.
89. Jaffee, David (2008). Organization Theory: Tension and Change. McGraw-Hill Education (UK) Ltd.
90. Jauch, Lawrence R. (1976). Relationships of Research and Teaching: Implications for Faculty Evaluation. Research in Higher Education. Vol. 5, No.1, 1–13.
91. Jick, Todd (1979). Mixing Qualitative and Quantitative Research Methods: Triangulation in Action. Administrative Science Quarterly, Vol. 24, No. 4, 602–611.
92. Johnson, D. (1994). Research Methods in Educational Management. Essex: Longman Group.
93. Keating, S. and Gabb, R. (2006). PBL in Engineering Student Expectations in 2006. Melbourne, Victoria University, Postcompulsory Education Centre.

94. Kerr, Klark (2001). The Uses of the University (5th edition). Massachuset: Harvard University Press.
95. Kirschner, Paul, A., Clark, Richard E. and Sweller, John (2006). Why Minimal Guidance during Instruction does not Work: An Analysis of the Failure of Constructivist, Discovery, Problem-based, Experiential, and Inquiry-based Teaching. Educational Psychologist, 41(2), 75–86.
96. Knelly, Leah (2006). The Development of a Problem-based Learning Classification Rubric for Community College Instruction. Unpublished master thesis. Oregon University.
97. Kolmos, A. and De Graaff, Erik (2007). The Process of Change to PBL. In Kolmos, A. and De Graaff, Erik (Eds.). Management of Change. Rotterdam: Sense Publishers. 31–44.
98. Kolmos, A., Fink, F. and Krog L. (2004). The Aalborg Model: Problem Based and Project Organized Learning. In Kolmos, A., Fink, F. and Krog L. (Eds.) The Aalborg PBL Model: Progress, Diversity and Challenges. Aalborg: Aalborg University Press. 9–18.
99. Kolmos, Anette (2002). Facilitating Change to a Problem-based Model. The International Journal for Academic Development. 7(1): 63–74.
100. Kolmos, Anette (2008). Problem-based and Project-based Learning: Iinstitutional and Global Change. In Skovsmose, Ole, Valero, Paola, and Christensen, Ole (Eds.). University Science and Mathematics Education in Transition. New York: Springer. 261–280.
101. Kolmos, Anette, and Algreen-Ussing, Helle (2001). Implementing a Problem-based and Project-organized Curriculum: a Cultural Change. Das Hochschulwesen, 1, 15–20.
102. Kotter, John (1996). Leading Change. Massachusetts: Harvard Business School Press.
103. Kotter, J. B. (1995). Why Transformation Efforts fail. Harvard Business Review. March-April 1995.
104. Kumar, Kamalesh and Thibodeaux, Mary (1990). Organizational Politics and Planned Organization Change: A Pragmatic Approach. Group and Organization Management, 357–365.
105. Li, H. and Henriksen, L. (2010). A Story of Organizational Change to PBL in Australia. Proceedings of the Joint International IGIP-SEFI Annual Conference 2010. European Society for Engineering Education.
106. Li, H. Du, X., and Stojcevski, A. (2009). How We Implement PBL in a University: Two Case Studies. In Gabb, Roger (Eds.). Proceedings of the 2nd International Research Symposium on PBL. Melbourne: Victoria University Press.

107. Lo, Yiuchun (1999). School-based Curriculum Development: the Hong Kong Experience. The Curriculum Journal. Vol.10, No.3, 419–442.
108. Lonka, K. (2001). The Helsinki 2000 curriculum: Challenges and Solutions in Transforming a Traditional Curriculum into Problem-based Learning. Paper presented at the Lernmodelle der ZukunftamBeispiel der Medizin, Munchen, DE.
109. Margetson, D. (1991). Why Is Problem Based Learning a Challenge? In D. Boud and G. Feletti (Eds.) The Challenge of Problem Based Learning. London: Kogan Page.
110. Marshall, M. (1996). Sampling for Qualitative Research. Vol. 13, No. 6, 522–525.
111. Martin, Thomas W. & Berry, K. J. (1969). The Teaching-Research Dilemma: Its Sources in the University Setting. The Journal of Higher Education. 40(9), 691–703.
112. Miles, M. and Huberman, A. (1994). Qualitative Data Analysis. London: Sage.
113. Mills, J. and Treagust, D. (2003). Engineering Education-Is Problem Based or Project Based Learning the Answer? Australia Journal of Engineering Education, (4).
114. Milne, M., Kostanski, M, Cara., C, Andison, K. (2008). Student Decision-making at Victoria University. Melbourne, Victoria University.
115. Moesby, E. (2006). Implementing Project Oriented and Problem Based Learning (POPBL) in Institutions or Sub-institutions. World transactions on engineering and technology education, 5(1): 45–52.
116. Moesby, E. (2004). Reflections on Making a Change towards Project Oriented and Problem Based Learning. World transactions on engineering and technology education, 3(2): 269–278.
117. Noor, K. (2008). Case Study: A Strategic Research Methodology. American Journal of Applied Science, 5(11): 1602–1604.
118. Norman, Geoffrey and Schmidt, Henk (2000). Effectiveness of Problem-based Learning Curricula: Theory, Practice and Paper Darts. Medical Education, Vol.34, 721–728.
119. Owens, Robert. G. (1970). Organizational Behavior in Education (2nd edition). Englewood Cliffs, N. J.: Prentice-Hall, c1970.
120. Owens, Robert. G. (2004). Organizational Behavior in Education: Adaptive Leadership and School Reform (2nd edition). Boston: Pearson, Allyn and Bacon.
121. Patton, M. (1987). How to Use Qualitative Methods in Evaluation. Newbury Park, CA: Sage.

122. Paul II, Chris W. & Rubin, Paul C. (1984). Teaching and Research: the Human Capital Paradigm. The Journal of Economic Education. Vol. 15, No. 2, 142–147.
123. Peter Schwartz, Stewart Mennin and Graham Webb (2001). Problem-Based Learning: Case Studies, Experience and Practice. London: Kogan Page.
124. Piaget, J. (1953). The Origins of Intelligence in Children. New York, NY: Basic Books.
125. Polkinghorne, Donald E. (2007). Validity Issues in Narrative Research. Qualitative Research, Vol. 13, No. 4, 471–486.
126. Powell, K. and Kalina, C. (2009). Cognitive and Social Constructivism: Developing Tools for an Effective Classroom. Education, Vol. 130, No. 2, 241–250.
127. Precision Consultancy (2007). Graduate Employability Skills-Prepared for the Business, Industry and Higher Education Collaboration Council.
128. Ramsden, P. (1992). Learning to teach in higher education. London: Routledge.
129. Savery, J. R. (2006). Overview of Problem-based Learning: Definitions and Distinctions. The Interdisciplinary Journal of Problem-based learning, Vol.1, No.1, 9–20.
130. Savery, J. R., & Duffy, T. M. (1994). Problem based learning: An Instructional Model and its Constructivist Framework. Educational Technology, 35(5), 31–38.
131. Savin-Baden, Maggi (2000). Problem-based learning in Higher Education: Untold Stories. Buckingham: Open University Press.
132. Savin-Baden, Maggi (2004). Challenging research in problem based learning. Berkshire: McGrawHill Education.
133. Savin-Baden, Maggi and Major, Claire. (2004). Foundations of Problem Based Learning. Berkshire: McGrawHill Education.
134. Savin-Baden, Maggi. (2003). Facilitating Problem-based Learning: Illuminating perspectives. The Society for Reseach into Higher Education and Open University Press.
135. Schön, Donald (1983). The Reflective Practitioner-How Professionals Think in Action. Basic Books, 1983. ISBN 0465068782
136. Schroeder, Ted. C. (1996). Integrating teaching and education. Journal of Agribusiness, 14(2), 129–142.
137. Scott, W. Richard (1987). Organizations: Rational, Natural, and the Open System (2nd edition). Englewood Cliffs, NJ: Prentice Hall.

138. Seawright, J. and Gerring, J. (2008). Case Selection Techniques in Case Study Research. Political Research Quarterly, Vol. 61, No. 2, 294–308.
139. Selznick, Philip (1996). Institutionalism: Old and New. Administrative Science Quarterly, Vol. 41, No.2, 270–277.
140. Senge, Paul (1990). The Fifth Discipline: The Art and Practice of the Learning Organization. New York: Doubleday.
141. Shenton, Andrew K. (2004). Strategies for Ensuring Trustworthiness in Qualitative Research Projects. Education for Information, 22, 63–75.
142. Shipman, H. and Duch, B. (2001). Problem Based Learning in Large and Very Large Classes. In Barbara J. Duch, Susan E. Groh, and Deborah E. Allen (Eds.). The Power of Problem-Based Learning (pp. 149–164). Sterling: Stylus Publishing.
143. Shor, Ira and Freire Paulo (1987). What is the "Dialogical Method" of Teaching. Journal of Education, Vol. 169, No. 3, 11–31.
144. Shuman, L. J., Besterfield, M., Mcgourty, J. (2005). The ABET "Professional Skills"- Can They Be Taught? Can They Be Assessed?. Journal of Engineering Education, (1), 41–55.
145. Simcock, A., Shi, J., and Thorn, R. (2009). Engaging Engineering Students with PBL (Problem Based Learning) Problems From Industry and Community. In: Innovations 2009: World Innovations in Engineering Education and Research. International Network for Engineering Education and Research (iNEER). Arlington, Va., USA. 273–285.
146. Smith, Page (1990). Killing the Spirit: Higher Education in America. New York: Viking Penguin Publishing.
147. Stojcevski, A., & Ozansoy, C. (2009). Problem Based Learning in Electrical and Electronic Engineering Education. In X. Du, E. de Graaff & A. Kolmos (Eds.), Research on PBL practice in engineering education (pp. 199–213). Rotterdam: Sense Publishers.
148. Strobel, J. and van Barneveld, A. (2009). When is PBL more effective? A Meta-synthesis of Meta-analyses Comparing PBL to Conventional Classrooms. The Interdisciplinary Journal of Problem-based learning. 3(1), 44–58.
149. Thinktask (2011). Three Years Ago. Retrieved from http://www.douban.com/group/topic/20301100/.
150. Trigwell, K., & Prosser, M. (1996). Changing Approaches to Teaching: A Relational Perspective. Studies in Higher Education, 21, 275–284.
151. Trigwell, K., Prosser, M. and Waterhouse, F. (1999). Relations between Teachers' Approaches to Teaching and Student Learning. Higher Education, 37, 57–70.

152. Turner, P. (1995). Student Learning in Electrical and Computer Systems Engineering. The 7th Annual Conference of the Australia Association of Engineering Education. Melbourne (10th-13th, December): 423–427.
153. Van Barneveld, A. & Strobel, J. (2009). Problem Based Learning: Effectiveness, Drivers, and Implementation Challeges. In X. Du, E. de Graaff & A. Kolmos (Eds.), Research on PBL practice in engineering education (pp. 35–44). Rotterdam: Sense Publishers.
154. Van de Ven, Andrew and Poole, Marshall Scott (1995). Explaining Development and Change in Organizations. Academy of Management Review, Vol.20, No.3, 510–540.
155. Victoria University (2007). Making VU 2016: A Statement of Purpose-Victoria University's Strategic Direction and Priorities 2008–2016.
156. Victoria University (2008). Victoria University International Profile.
157. Victoria University (2009 a). About VU. Retrieved from http://www.vu.edu.au/about-vu.
158. Victoria University (2009 b). Campuses. Retrieved from http://www.vu.edu.au/campuses.
159. Vijava, T. And Hemamalini, R. (2012). Role Ambiguity, Role Conflict and Work Role Balance: Influence on Organizational Commitment and Turnover Intention of Faculty. European Journal of Social Science. Vol. 30, No. 4.574–585.
160. Vygotsky, L. S. (1962). Thought and Language. Cambridge, MA: MIT Press (original work published in 1934)
161. Walton, H. J. and Mathews, M. B. (1989). Essentials of Problem Based Learning. Medical Education, 23: 542–558.
162. Wang, G., Tai B., and Huang, C. et al. (2008). Establishing a Multidisciplinary PBL Curriculum in the School of Stomatology at Wuhan University. International Dental Education, 72(5): 610–615.
163. Weick, K. E. (1976). Educational Organizations as Loosely Coupled Systems. Administrative Science Quarterly, 21, 1–19.
164. Wengraf, Tom (2001). Qualitative Research Interviewing. London: Sage Publications.
165. Yin, Robert. (1994). Case Study Research: Design and Method (2nd Edition). Applied Social Research Methods Series, 5. Thousand Oaks, CA: Sage.
166. Young, Pat. (2006). Out of balance: Lecturers' Perceptions of Differential Status and Rewards in Relation to Teaching and Research. Teaching in Higher Education. Vol. 11, No. 2, 191–202.

Index

For Product Safety Concerns and Information please contact our EU representative GPSR@taylorandfrancis.com
Taylor & Francis Verlag GmbH, Kaufingerstraße 24, 80331 München, Germany

www.ingramcontent.com/pod-product-compliance
Lightning Source LLC
LaVergne TN
LVHW010627300926
849162LV00001B/86

* 9 7 8 8 7 9 2 9 8 2 7 0 4 *